SACRED GEOMETRY IN ANCIENT GODDESS CULTURES

"Richard Heath's fascinating and highly readable book presents a decoding of the metrology of Göbekli Tepe of 10,000 BCE and advances arguments that support its role as a place for the worship of the Goddess. Although the megaliths of Europe were to come nearly 5,000 years later, the book marshals evidence from mathematics, astronomy, sacred architecture, and tone theory to show that these structures were a successor to Göbekli Tepe. The book disentangles many threads that went into the creation of the Classical Age."

SUBHASH KAK, REGENTS PROFESSOR AT OKLAHOMA STATE UNIVERSITY
AND AUTHOR OF *THE ASTRONOMICAL CODE OF THE ṚGVEDA*

"Sacred numbers have always been a part of Richard Heath's writing, and it is no surprise that in this book he makes the case that religion developed from a pure science of horizon astronomy. This ancient knowledge evolved over thousands of years and influenced religion as practiced today. The author describes three major themes: that horizon astronomy developed in matriarchal societies thousands of years ago; that this knowledge was subsumed by a transition to a patriarchal system; and that in this transition, numbers—having lost their true meaning—became sacred. You may be left pondering what if the transition from Goddess to Father had not taken place."

DAN PALMATEER, ANCIENT GEOMETRY RESEARCHER

"The esoteric knowledge encoded in ancient sacred sites is being revealed now. As the result of more than two hundred years of global archaeology and site analysis using metrology, sacred geometry, and cosmological analysis, surprising insights about sacred places are coming forth. These remarkable findings are brilliantly described in Richard Heath's *Sacred Geometry in Ancient Goddess Cultures*. He shows us how Mesolithic priestesses and Neolithic priests designed sacred places to resonate with the powers of Earth in synchronicity with the cycles of the moon, planets, and stars. Then later cultures, such as the Greeks, followed ancient guidance as well as using music, harmony, mythology, and storytelling. As I contemplated these insights pouring out of Heath's marvelously insightful mind, I felt a bond with these creators; I could *feel* their joy! This wonderful and beautifully illustrated book invites us to feel Gaia, the sacred feminine of Earth. For me, this ancient lineage can be described as the Matriarchy through Patriarchy into Humanarchy—our return to ecological worship—knowing Earth in our bodies as we contemplate the cosmos."

BARBARA HAND CLOW,
AUTHOR OF *AWAKENING THE PLANETARY MIND* AND
REVELATIONS FROM THE SOURCE

"A valuable body of research in which Heath weaves a fine web of intrigue and illumination."

MARK VIDLER, COAUTHOR OF *SACRED GEOMETRY OF THE EARTH*

SACRED GEOMETRY
IN ANCIENT
GODDESS CULTURES

The Divine Science of the
Female Priesthood

RICHARD HEATH

Inner Traditions
Rochester, Vermont

Inner Traditions
One Park Street
Rochester, Vermont 05767
www.InnerTraditions.com

Cataloging-in-Publication Data for this title is available from the Library of Congress

ISBN 978-1-64411-655-5 (print)
ISBN 978-1-64411-654-8 (ebook)

Printed and bound in China by Reliance Printing Co., Ltd.

10 9 8 7 6 5 4 3 2 1

Text design and layout by Debbie Glogover
This book was typeset in Garamond Premier Pro with Cantoria MT Std, and Gill Sans MT Pro used as display typefaces

To send correspondence to the author of this book, mail a first-class letter to the author c/o Inner Traditions • Bear & Company, One Park Street, Rochester, VT 05767, and we will forward the communication, or contact the author directly at **sacrednumber@gmail.com** or through **sacred.numbersciences.org**.

Scan the QR code and save 25% at InnerTraditions.com. Browse over 2,000 titles on spirituality, the occult, ancient mysteries, new science, holistic health, and natural medicine.

CONTENTS

PART THREE

THE ROLE OF PROVIDENTIAL HISTORY

Explicatory Poem

Who did spin this wheel of meaning!
Why did my language make this reading?
Was it just my own interpretation of,
the ancient mind, its shackles off?
Or some desire, to understand
the form of history, as if grand?
Or could it be, the geocentric planets,
our old gods, sponsored my rackets?
Or did all of these conspire to touch
the Talisman that makes so much?
A vision in the longer darkness,
unseen by those who live in lightness.

The Three Streams

Whereas seven is greater than twelve,
and three greater than seven.
One is greater than all these;
But eight is still greater than four,
and four greater than two.
while anything with five in it,
is the same.

PREFACE

Over many years I searched for where myth and science join. . . . Number gave the key. Way back in time, before writing was even invented, it was measures and counting that provided the armature, the frame on which the rich texture of real myth was to grow.

GIORGIO DE SANTILLANA AND HERTHA VON DECHEND,
HAMLET'S MILL

This interpretation of human history since the Ice Age began with the Mesolithic monument called Göbekli Tepe, built just after the ice had receded. Having challenged an email correspondent* that if the foot-based metrology used in the megaliths was employed at Göbekli Tepe, 5000 years before the megalithic age began, then the megalithic could not have originated that technology, which was mankind's *only* metrological scheme until the French meter. The English foot originated to enable pre-numerate astronomers to do calculations based on whole-number fractions of that foot, and, at Göbekli Tepe, a subunit of measure (12/77 feet) was evolved from that foot, having unique and useful characteristics regarding π. This subunit was invaluable when building stone structures involving circles, because it divided well into important variations of the foot called Sumerian (12/11 feet), Royal (8/7 feet), and Egyptian (48/49 feet). I soon found the subunit elsewhere in the eastern Mediterranean region on the megalithic island of Malta and then in the early Christian rock-cut churches of Cappadocia.

The matriarchal cultures of the southern Mediterranean seem to have had a special relationship linking the megalithic astronomy in Malta to the strange

---

*Fred Morris Jr.

T-pillared stone circles of Göbekli Tepe. And after the significant gender transition, at the end of the Greek Bronze Age, there was notable continuity of matriarchal cultural forms within the patriarchal tribes that then forged the classical world, following the assimilation of the Minoans and the mainland matrilineal tribes. Later again, the Romans assimilated the upland Etruscans, who were matriarchal, before establishing their empire throughout Europe.

I began to question whether the coastal megaliths of fifth-millennium Europe could also have been built by Middle Stone Age (Mesolithic) tribes of foragers, who were generally matrilineal. It has become normal to call the megalithic the work of a Neolithic culture, who are almost invariably patriarchal farmers adopting a well-defined "Neolithic package" of pastoralism. The Neolithic had developed in the Near East, south and southwest of Göbekli Tepe. It was sweeping through *central* Europe's Mesolithic populations around the same period as the megaliths were built on the Atlantic coast. In which case, why should the Göbekli Tepe stone circles appear in 10,000 BCE with a form of megalithic stone enclosure employing similar fractions of the English foot?* After all, for the builders of Göbekli Tepe the Ice Age had only just ended, and the Neolithic only just beginning to develop farming techniques.

From later Vedic texts, one can see a new type of descriptive language had been developed that became a template for the development of all later Indo-European languages in new Neolithic regions. The early Vedas are undoubtedly astronomical yet strangely obscure. One fact that may explain this is that, during the late Ice Age, the western Arctic Sea and its coasts were warmed by the Gulf Stream, creating an interglacial warming period lasting at least two millennia. The Indian scholar B. G. Tilak demonstrated that a major use for this language was to describe, in mythical terms, the peculiar astronomy seen only in the sky at the North Pole, described in the Vedas as the holy mountain Meru, and related names in the later Indo-European languages. And J. G. Bennett saw that astronomy of the Arctic was advantageous to the project of language creation to *describe a large and coherent phenomenon,* which polar astronomy uniquely is, and in some ways more holistic than the astronomy at lower latitudes upon the Earth.

The builders of Göbekli Tepe could well have migrated from this Arctic Center, while Europe's later megaliths, of the fifth millennium, were a redoing of the same type of astronomy but at lower latitudes. But Mesolithic tribes

---

*Not made in England but certainly in-use in the British megalithic.

would have needed the Arctic grammar to describe and remember the astronomy, in spoken form and through the geometrical Art of the Goddess. Before modern astronomy, events on the horizon enabled the geocentric planetary world surrounding Earth to be discovered as based upon invariances within the number field and through simple geometrical forms. Göbekli Tepe is especially distinguished in being located at the epicenter of an ideal fertile region for agriculture through which, over thousands of years, the Neolithic revolution was developed, and, as the Neolithic diffused through Europe, it carried a transforming Indo-European speech into a large range of Mesolithic dialects.

With farming also came the patrilineal way of life, and matriarchy was subsumed. The new patriarchal gods and myths became so similar to those of the Vedas because of this diffusion of the Proto-Indo-European language, with the Neolithic. The languages of European matriarchal tribes became Indo-European as patriarchal farming was adopted, in a uniquely significantly stage of human history, leading to the fusion of (a) the astronomical methods and data of the Megalithic and, (b) the new mathematical thinking of patriarchal Greece and the ancient Near East. Literacy in Classical Greece combined alphabetic writing with the new grammar, to write and read descriptively, taking snapshots of the ancient myths and cosmologies, while innovating a new breed of speculative philosophies.

In my view, therefore, the Neolithic slowly developed in parallel with a pre-existing Mesolithic way of life organized around women and their children. The eventual Neolithic pastoralists were instead organized around men, their farms, family, and roles in society. The Current Era leading to Western civilization was, in significant part, the result of these two streams having been kept apart and intact, until the diffusion of the Neolithic package in Greece only latterly swept away the Mesolithic way of life, while dramatizing it in Greek myths and many other institutions, as the cultural traces of the matrilineal past we have today.

The megaliths of the Atlantic and Mediterranean coasts then appear to *recapitulate* Göbekli's Tepe's geocentric astronomy of the Arctic center, and, because of improved climate during the Holocene maximum, Mesolithic peoples could re-express their "hobbies" of horizon astronomy, time counting, and geometrical expression of the sophisticated pattern of time surrounding the Earth. The monuments of an Arctic center, lost in the terminal Ice Age maximum, would have evolved similar buildings and tools (metrology) and techniques (day-inch counting) as those used to build the later megaliths.

All metrology on the Earth has had a single source, related to measuring the heavens and the earth. The Arctic foot-based metrology, visible at Göbekli Tepe, must have been transmitted to the later megalithic builders, who had the root foot as a standard. An oral tradition would have been necessary, using a descriptive type of language seen in the Vedas. The standardized English foot was the number one for the system. And it was probably the Arctic center that led to the Atlantis myth, of a lost center of high knowledge.

The megalithic period naturally leads one to question why humans saw, in astronomy, a means to discover "in the first Place" the purpose of themselves and their world. This astronomy was evidently seen as having a religious or heavenly dimension, now discarded and altogether invisible to the modern norms of irreligion and an abstracted heliocentric science of physical laws as the only causation of the world.

PART ONE

MEGALITHS
BUILT
BY THE
GODDESS

Lift up thine eyes round about, and see.

ISAIAH 60:4

The heavens declare the glory of God;
and the firmament sheweth his handiwork.

PSALM 19:1

If the New Stone Age (or Neolithic), 9000 BCE to 3000 BCE, was an age chiefly defined by farming technologies (from the Near East), this conflicts with the dates for the megalithic period (in northwestern Europe), which was not about farming. Megaliths were typically built on the marine hinterlands of the Mediterranean and Atlantic coasts, between 5000 BCE and 3000 BCE. At this time, the Neolithic advances in agricultural technologies were only starting to make inroads into the lives of people living in most of Europe—in regions that had not yet undergone the centuries of clearance, improvement, and adaptation needed and thus were better suited to foraging rather than farming. In contrast, the Near Eastern civilizations had arisen upon a land and climate with native species ideal for farming and domestication. The seed stock of existing grasses and pulses could be selectively improved from the flora growing wild as ground cover, and the indigenous animal species soon became domesticated. By 5000 BCE what had been learned about agriculture could be applied to the already rich soil in Mesopotamia, using irrigation from its two rivers. But the impulse toward farming does not naturally lead to the nonproductive—in terms of survival, that is—activity of building large stone structures with astronomical functions.

And this fact is now clear: the megaliths were associated with some form of practical astronomy. It has been argued that the primitive calendars based on the lunar month were superseded by or integrated with later solar calendars developed to organize the seasonal life of the farmers. But the megaliths were rarely located in naturally pastoral settings, and the knowledge of the Sun and Moon they enabled appears to have been a project of knowledge rather than of providing farmers with a seasonal calendar to replace the earlier lunar ones.

Near Eastern civilization only appeared, in Sumeria and Egypt, as the megalithic was ending, and its ancient temples and religious buildings appear to offer evidence for a continuity of megalithic traditions by their use of geometric templates incorporating symbolic geometries and the numerical science of measures evolved by the megalithic astronomers and first developed within megalithic buildings. Therefore, one must conclude that megalithic standards came to form a basis for high-status religious and royal building made possible through the excess produce provided to cities by their surrounding Neolithic settlements. The temples built in the cities were based on the discoveries made by the megalithic builders, and they were largely drawing on astronomical symbolism.

In fact, megalithic buildings are unlikely to be related to Neolithic farming at all. Instead they belonged to the people of the Middle Stone Age (the Mesolithic), the term used for the *state* the peoples of Old Europe were in before they settled into the lifestyle called Neolithic, *even if* they were living in the same epoch as the Neolithic movement that was expanding west from the Near East. When the last Ice Age ended, the Mesolithic and Neolithic—or New Stone Age—predated the Metal Ages of the historical period; that is, there was a sizeable geographical overlap of two completely different ways of living, as the Neolithic way only slowly progressed through Europe, over one or two millennia. Confusion has arisen, owing to archaeologists automatically attributing Mesolithic buildings to the Neolithic. Only very recently have the two streams of megalithic and Neolithic been teased apart in time and space on the map of Europe, revealing the Mesolithic transformation into the Neolithic occurring over millennia.*

This overlap and confusion of lifestyles illustrates how historical movements interact and take on a historical, geographical *form*. From these forms deductions can be drawn, revealing more than the factual evidence generated by the carbon and other dating methods, layers of deposits, connected artifacts, and so on, of archaeology. Accepting such deductions requires the visualization of influences within time, without any glib assumption that the megalithic belonged to the New Stone Age, for instance. The twin movements of farming and sky watching were contemporary but geographically exclusive, and their life ways were very different. Once one takes this step of seeing the megalithic as a manifestation of the Mesolithic, then one can see the matrilineal social organization common to the Mesolithic. Matrilineal tribes were very large extended families, and so they were able to build stone monuments because their food was provided from the wild and (in the presence of the kinder conditions after the Ice Age) there was time to build megaliths to study astronomical time.

Being focused on the woman's role and her fertility, the natural religious instinct for the Mesolithic was the idea of God as Mother, who was the source of the world and "all that lives," which is the meaning of the name "Eve" in the

---

*It was Barry Cunliffe in 2008 who did the mapping and Bettina Shulz Paulsson in 2017 who filtered out less reliable carbon dates for the megalithic to establish the earliest dates that can be trusted.

Bible. The art of the female goddess had, for many millennia, expressed a wide-ranging symbolism that was interwoven with what could be made of astronomical time and the visual hierarchies seen in the sky itself.

THE ATLANTIC MEGALITHIC

This goddess symbolism is very clear in the matriarchal art found near the oldest Atlantic sites, such as those around Carnac in Brittany. Four kilometers east of the village of Carnac, a chambered corridor cairn called Gavrinis was built around 3500 BCE. The passage was organized to shutter the light from the solar and lunar extremes on the horizon to the southeast onto the third central "end stone," now called C3. These two types of extreme events do not occur simultaneously, and the design was cleverly based on the diagonal of a 7-square rectangle within the corridor's design; this rectangle's diagonal generates the angle between the extreme Sun and Moon at the latitude of Carnac (see fig. P1.1).

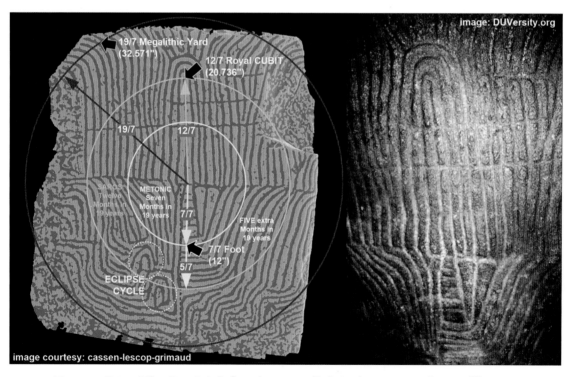

FIGURE P1.1. The Gavrinis left end stone called C3 (*C* meaning "central") upon which the light shines, (*a*) of the sun at the winter solstice sunrise, and (*b*) at the full moonrise during its southerly maximum standstill, every 18.618 years.

To form the corridor to the chamber, already existing engraved stones* had been taken from megalithic sites near Carnac. These designs, pecked on granite, display a technical language to express how time emanates from horizon events, like ripples in water. Each wave is a unit of measure, such as the day represented as an inch between the emanating waves as when a stone is dropped into a still lake. The waves therefore show the numbers of days, months, and years between events, counting the time between astronomical events and these alignments.

The engraved stones of the left and right walls of the corridor were often divided vertically into three regions with a base, middle, and top panel (see fig. P1.2), and these can perhaps be interpreted as representing different aspects of a given astronomical phenomenon: how the phenomenon looked, how the numbers looked, and the different measures required to capture and represent it. This extends the usefulness of such a quasi-geometric language, seen also in the Irish cairns such as Newgrange, where a part of an actual megalith found at Carnac has been found embedded within Newgrange, alongside motifs similar to the art of other goddess cultures. When Gavrinis was no longer needed it

FIGURE P1.2. Stone R8 documented the Metonic and Saros periods of 19 solar and 19 eclipse years.

---

*The stones have signs of weathering from different directions.

was backfilled with sand, only to be excavated in the nineteenth century. This covering over of an intact monument after its usefulness ended was also seen at Göbekli Tepe, and although a natural phenomenon such as a tsunami could also provide an explanation for its burial, such as at Scara Brae in Orkney, this was not the case here. Closed by 3000 BCE, the survival of these decorated stones makes Gavrinis one of the wonders of prehistory. The stones give a final report on how sophisticated the megalithic enterprise was around Carnac—the earliest in-monument carbon dating being in the closed crypt of the Saint-Michel tumulus nearby—as about 5000 BC.[1]

No signs of settled habitation have been found in the Bay of Morbihan, perhaps because the construction work was done by a band of foragers living the Mesolithic lifestyle. The geographically close work in Ireland had already begun and, across the Irish Sea, in western Wales, a stone circle built from bluestones quarried in the Preseli Hills became Stonehenge 1.[2] Chapter 3 proposes that the Atlantic branch of the megalithic was further developed at Stonehenge 3 to include contributions from the Mediterranean and Near Eastern branches of the megalithic "franchise," in a reunion that built this most memorable yet idiosyncratic stone circle, where both a geodetic model of the Earth and a time model of Venus were added to the Welsh bluestone monument, using then the plain but dressed local sandstones, called Sarsens, of the later Stonehenge 3.

THE MEDITERRANEAN MEGALITHIC

The most singular megalithic buildings in the Mediterranean are on the archipelago of Malta. These clearly express the mother goddess, giving support to the notion that the megalithic had been a matriarchal phenomenon of post–Ice Age Mesolithic people. Malta's megalithic astronomy is distinguished by its latitude of 36 degrees north, a latitude from which emanated the star constellations reported by Eudoxus that have recently been shown to exclude those stars *not visible* from 36 degrees north around 2800 BCE, during what was the megalithic period on the archipelago. From this, we see that the outlook south over the Mediterranean influenced the focus for Malta's megalithic astronomy, naturally differentiating it from the higher latitudes of the Atlantic Coast, where the systematic study of constellations was more difficult but where the circumpolar astronomy becomes ever clearer, leading to the radical astronomy of the Arctic Circle, with its single day and night per year (see chapter 8).

Another key significance of Malta was its relation to the larger island of Crete, 800 miles to the east. In the second millennium BCE, Crete became highly influential on account of its maritime trade, but the values of its matriarchal *civilization* would increasingly clash with those of the patriarchal Neolithic tribes then entering mainland Greece from the north and east, probably as the result of worsening weather in the north or the population pressure caused by Neolithic farming, which both drove populations to the south. Malta and Crete were close to Greece and the Greek invention of writing, literacy, and the same kind of written history invented by the writers of the Hebrew Bible, whose writing was made possible by the invention of the phonetic alphabet. Perhaps the confrontation of these two different civilizations is what the "clash of the Titans" refers to: the time when a Mesolithic planetary god (Cronos/Saturn) was displaced by Zeus/Jupiter, resulting in the patriarchal towns and cities of the past two millennia up to the present, a conflict glossed over by classical Greek myths, the history of the Roman Empire, and the Bible.

The tendency to isolate regions from each other—as is the case with megaliths—and then always connecting them to the Neolithic has inhibited our understanding of their history. Our story here has been aided by studying the significance, among megalithic peoples, of whole numbers and geometry in the conduct and context of their astronomy and the subsequent transmission of exact measures and cosmic stories based upon astronomical invariance. This is another *structural form* of megalithic techniques involving number but which is ignored by our histories. The special study of megalithic astronomy, still denied as a subject of study by many specialists, has enabled this story to be recovered; it will be seen to complement the official story and the conclusions offered by our modern specialists.

1

THE LANGUAGE OF
THE MESOLITHIC

The role of women in society has changed since the Bronze Age from one of being central to the life of tribes to being associated with patriarchs, predominantly as the wife of a farmer or an urban specialist. The pastoralism of the Neolithic could only be portable when settling regions good for arable and livestock farming. Any Neolithic diffusion into Europe demanded adequate soil and weather suitable for raising the flora and fauna carried along with prospectors, both now improved by selective breeding in the fertile crescent. Neolithic methods would be more marginal in many regions, so these were skipped over, and it was the Great Hungarian Plain that became the chief conduit west. The Near East and Nile Valley were ideal for agriculture and stand out in our histories as cradles of civilization. The Fertile Crescent's ideal locations became the source of a Neolithic diaspora that pushed into central Europe, and the diffusion of this new way to live was largely overland and at higher latitudes rather than along the Mediterranean. The west was meanwhile occupied by the Mesolithic peoples of Old Europe, and a given region of good land could take centuries to become properly settled.

THE NORTHERN AND SOUTHERN BRANCHES
OF MEGALITHISM

The Neolithic journey to the west moved through Asia Minor (Anatolia), crossed the Aegean Sea, and then headed north through mainland Greece, whose coastal plains had been greatly reduced by the inundation of Ice Age meltwaters. Beyond the mountains to the north lay present-day Bulgaria and the

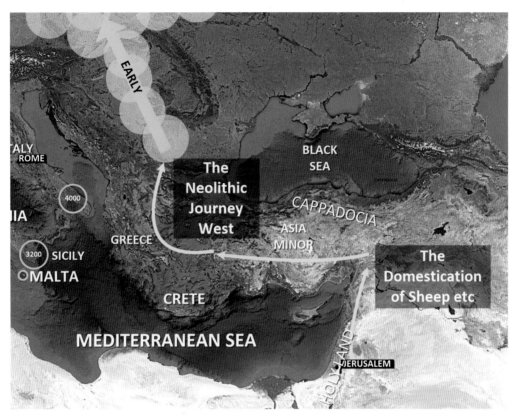

FIGURE 1.1. *Dated circles,* the dating of megalithic sites of the Atlantic and Mediterranean. From Paulssen, *Time and Stone,* 5. *Colored circles,* the diffusion of the Neolithic package from the Balkans to the Paris Basin via the Great Hungarian Plain. From Cunliffe, *Europe between the Oceans.*

Great Hungarian Plain leading to northwestern Europe. We see the Neolithic arriving in the region of the Paris Basin only around 5000 BCE.[1] It was about this time that the megaliths of Carnac appeared in the far west of France, in Brittany. The megaliths were probably not built by the migrating carriers of the Neolithic package but rather by the indigenous Mesolithic peoples organized as matrilineal tribes and able to provide a large and reliable workforce supported by the foraging self-sufficiency learned in the Ice Age (see fig. 1.1).

From their initial center in Brittany, the monuments were especially concentrated around the Morbihan (Carnac). This tradition, judging by its architecture and art, later moved to Ireland and then to the megalithic area of western Wales, epitomized by Pentre Ifan, the largest portal dolmen in Europe. From there, many bluestones from the Preseli Hills were transported to southern England to create the outer monument of Stonehenge, a stone circle of fifty-six stones. The site of Stonehenge was placed one-quarter of a degree in latitude from the very large henge at Avebury* (see chapter 3), around 3000 BCE. A continuous granite massif runs under the sea between Carnac and the Preseli Hills.

Around the same time (5000 BCE[2]), there was another center of megalithism, in the western Mediterranean on the island of Sardinia and northwestern Spain. This worked its way east, along the coast or by island hopping toward Greece, through lands occupied by Mesolithic tribes. There was limited migration of the Neolithic settlers from the east, for want of good farming lands.

STEPPING-STONES TO NUMERACY

Once one talks about the Stone Age measuring angles, or lengths, one must account for the fact that measurement involves units of measure and that measurements are numbers. The megalithic mind may have already developed geometry within its art, using measured (or measurable) lines or radii. Exact and repeatable units of length can both make and measure geometrical patterns. But units of length can also make linear measurements of time counted *as exact lengths* so as to know and predict events in the sky.

Drawing geometrical shapes can reveal the abstract properties of small numbers when using constant units of length: for example, the numbers 3 and 4 can be represented by lines 3 and 4 identical units long; if these lines are then placed at right angles to each other, they create a special rectangle (see fig. 1.2). The diagonal between the corners is then automatically 5 units.

Figure 1.2 shows that using multiple identical squares within the construction process to make a rectangle with sides measuring 3 and 4 produces a diagonal made of 5 units. Long before Pythagoras, this simplest of the whole-number Pythagorean triangles was easy to find in this way. The next most simple triangle has sides of {5, 12, 13}[†] units long. This whole-number triangle is found alongside the earliest astronomy of the megalithic where alignments along the ground (leading to the horizon), located celestial events (on the horizon),

---

*Avebury, is henge with a surrounding ditch and outer bank that has a flat interior space in which stone circles of various sorts were later articulated, perhaps when stone circles started to appear after the Cursus culture of earthworks.

†I will sometimes use the curly braces of set notation to denote sets of integers, or other things. {3 4 5} is the same as 3-4-5 and 3, 4, 5.

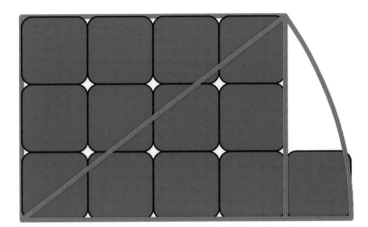

Figure 1.2. The numbers 3 and 4 as lengths in units generate the next integer, 5, as the diagonal or hypotenuse of the right triangle.

then counted as the length of the days in between events.. But in practice the Mediterranean could avoid the difficulty of forming a triangle by making a rectangle, in this case of 3-by-4 units (see, for instance, the Antioch mosaic in figure 5.16 on p. 103). There are two identical but contraflowing triangles in every rectangle. So while we see a triangle at Le Manio, near Carnac, the more holistic form is a rectangle of the two smaller sides. The rectangle's two equal diagonals are the hypotenuse—which in Greek means that which is "stretching under" the right angles, like the rectangle's diagonals.

Integer lengths, the rational fractions between rectangular sides and a diagonal, or a circle's radius to its circumference, are a different language from the decimal fractions in our base-10 notation or any other positional numbering system of the Near East in which numbers less than one form fractional parts of powers of 10 or 60. In the past, numbers less than 1 were represented as rational (integer) fractions (such as ¾), which enabled smallness to be seen as symmetrical to largeness (as with 4/3 being greater than 1, versus ¾ being less than 1). But keep in mind that counted lengths as measurements were *baseless,* being just the ordinal numbers of a measuring unit, with aggregated sums such as 59 units being equal in days to two lunar months, when counted.

In ancient-world cosmologies the number 1 was often—as it was in metrology—considered central, it being the cause of all numbers because each integer number (seen one way) is a number of ones, and all the rational fractions involve division, as with 9/8, where a whole number 9 is divided by eighths. Fractions greater and lesser than 1 formed a pair of "twins," geometrically symmetrical to the number 1, or unity, as, for example, the set {8/9, 1, 9/8}. Functional arithmetic displaced this older symmetry of integers, and natural triangles such as the 3-4-5 lost much of their significance for demonstrating that *integer numbers govern how things fit together geometrically* in the actual world of space and time. Things are unities in themselves, but a number of them, when measured, is an integer measure. From this, a subunit like the inch could count days and record, as a length, what we call a natural number that was equivalent to a length of time in days. We still say, "How long will it take?"

The prehistoric world therefore came to the idea of space and time as a *counted* length and used this to study astronomical time; this became the fundamental sense of how the celestial bodies uniquely generate time within our earthly experience through their rotations (like the day), orbits (like the year), and ties between anomalous retrograde motions of planets called synodic periods. Superficially, time is just night and day, the lunar phases, or how many of these divide into the solar year. But ancient studies of the night sky showed that the "wandering" planets expressed ratios of one to another that, when expressed as lengths (in space) by integer counts of whole days or months (in time), could break out of the single dimension of time counting and create right triangles (or rectangles) to compare lengths or circles to simulate time as a reentrant cyclic structure as in the symbol of the Ouroboros snake* that eats its own tail. A new discipline of *ratio management* emerged, requiring metrology to be invented as a science of measures in which right triangles with integer (longer) sides became proportional calculators between numbers as length, that are lines. These triangles were soon superseded by the fractional foot measures themselves that must be generated using right triangles, from a single standard length, with the foot as the number 1 (in the ancient foot-based metrology). Fractional measures such as 9/8 feet could be created using a triangle of base 8 feet and diagonal 9 feet. Above each foot on the base of 8 feet was a foot on the hypotenuse of 9/8 feet.

Time was also related to walking, so that a standardized foot became, for the purposes of calculation, the measure of one unit in motion, a pace equaling 2.5 feet, and a step of two paces was 5 feet, and so on. The earliest experiments along this line of inquiry evidently used the inch (a thumb's width) or a digit (a finger's width) to count days. But it was soon realized that a right-angled triangle (also called a right triangle), when used for comparing two counts, could perform proportional *calculations* in feet of 12 inches. Once the larger foot unit was adopted, different foot ratios could form the basis of a whole set of derived measures (or foot modules, like the Royal foot of 8/7 feet) to both exploit and

---

*As in the Ouroboros snake who eats its own tail as a symbol of a reentrant cycle of time. In Greek myth, characters with tails rather than legs were used to signify celestial entities.

explore the properties of the natural numbers and to calculate, well before the age of numbers as symbols emerged and with it the manipulation of numbers using arithmetic.

The domain of the natural numbers has led to the modern discipline called number theory, but the megalithic astronomers were talented and studious pragmatists rather than theorists. Remember that theory, in history, was only resolved by classical Greece. Before that, there were methods to do things.

Based upon fractional feet, this pre-arithmetic numeracy continued to be used within ancient buildings, but its role in understanding the time patterns of eternity had become symbolic at best, say, to the master masons building Chartres Cathedral, for example. An "eternity blindness"* had set in, *inhibiting* the perception of what is, and must have been, the original usefulness of metrology for astronomical calculation.

THE GEOMETRICAL ORIGINS OF THE MEASURES

As stated above, the tribes of the eastern Mediterranean basin were, like Old Europe, originally organized in matrilineal fashion: the women were mothers, raising children who were given the name of a significant matrilineal ancestor,† while the menfolk were uncles to all the tribe's children. A matrilineal tribal organization probably arose because of living communally for survival, since it was women who created the tribe's members and, having raised them, such Mesolithic tribes could then forage as a team. After the Ice Age, the decrease in pressure for survival left time for Mesolithic tribes to engage in the large megalithic project of astronomy, prefigured for us by Göbekli Tepe. The Mediterranean still retains aspects of these matrilineal societies despite the victory of patrilineal pastoralists. As stated above, patriarchy seems to have arisen naturally from the Middle Eastern Neolithic package, which led to urban development and population growth based on the new economy of farm-supplied sustenance. If civilization was made possible by the Neolithic, then the matrilineal

---

*An important term of mathematician and natural philosopher J. G. Bennett, who defined a dimensional framework for the universe in which time and space (now linked by relativity and called space-time) was one of three types of time-and-space-like dimensions, one of which is our space-time, another the eternal recurrence of the planetary system, and another the redemptive time in which *irreversible creative* steps change the world as to its inevitable fulfillment, due to an act of will rather than of being.

†Similar to the way the later Hebrew tribes used the masculine name "the children of Israel."

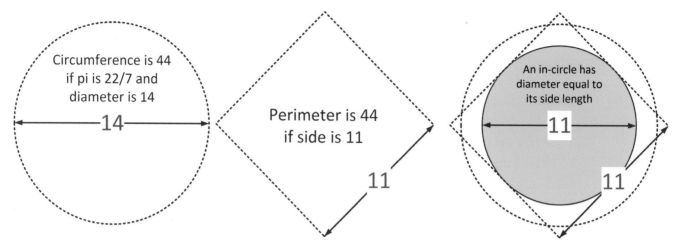

FIGURE 1.3. A circle equal in circumference to the perimeter of a square will have an in-circle diameter equal to the side of the square.

tribes were probably the unreconstructed populations of the Ice Age foragers in Europe, who then built Göbekli Tepe. Maternal roots for a society would have been reflected in the way sacred geometry was conceived, perhaps as a process of birthing, a coming-into-being of new relationships and forms.

The rotund form of the mother goddess is governed by π. Measures containing 7 and 11 could exploit one of the earliest approximations of π* as 22/7, which is in turn strongly associated with the geometry in which a square is constructed of equal perimeter length 44 to a circle drawn concentrically to the square of diameter 14 (see fig. 1.3).

Perhaps the accurate π of 22/7 became clear through trial and error: to adjust whole-unit diameters, 7 in the radius results in 4 × 11, equal to 44, in a circle's circumference. This means that a square with side length 11 would have the same perimeter as a circle of diameter 14, both being 44 if π is 22/7. A square of side 11 has, in turn, an in-circle of diameter 11. It is clear that the four quadrants of a circle radius 7 must each be the same length as each side of a square whose side length is 11 and perimeter 44. This is how the megalithic peoples could have, using small numbers, come across this remarkable geometry of equal perimeters for the circle of diameter 14 and square of side 11 (see fig. 1.4).

This numerical geometry had many practical applications and meanings. If

---

*The ratio between the length of a circle's diameter to its circumference.

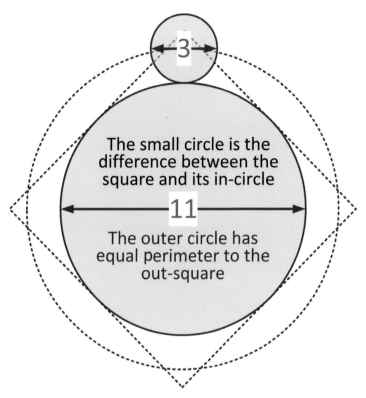

FIGURE 1.4. The difference between circles of 14 and 11
gives the difference between 4 and π and the radius 1.5:
14 minus 11 equals 3 so that the circle of 14 exceeds the square of
11 by 1.5 on each side, as radius of the small "lunar" circle.

you take a circle with 360 degrees marked on the circumference, a square of side
length 90 could be made, and a circle of diameter 14/11 × 90 would then have a
circumference of 360 feet since 14/11 × 22/7 = 4. In terms of English feet, 14/11
is to be found as a known unit of measure,* so that 90 feet of 14/11 feet would be,
by ancient times, the way to make a circumference with a radius of 45 × 14/11 feet
to make a circumference of 360 that, from the center, would be spaced by one
Sumerian degree per foot on the circumference. Thus, 360 degrees were chosen
for a circle's inner metric from the center because of the transmission from linear
measure (the diameter) to the angular measure from the center (8 × 45 = 360,
half of 720).

---

*A *remen*, or 6/5 (1.2), of the inverse Samian foot, equal to 35/33 feet. The 14/11-foot subunit
emerges in the Haçlı Kelise rock-cut church in Cappadocia (see chapter 7).

The smaller in-circle of diameter 11 is cosmically significant: if the Earth is equated with the circle of side 11 and the Moon equated to the circle of diameter 3, 11 to 3 is the correct ratio of the diameter of the Earth to the diameter of the Moon. The implications are that a cosmic intelligence, associated with the solar system, chose to give the Earth and Moon this simple ratio, which emerges from the equal-perimeter model when π is, most simply, 22/7. Its cosmic profundity is even clearer when the dimensions are multiplied by 720, itself an important limiting number* in musical harmony. Thus, 720 is the multiplication of the first six numbers {1, 2, 3, 4, 5, 6}; that is, the megalithic seem to have received or worked out a harmonic model in which the ratios between the first six numbers {1, 2, 3, 4, 5, 6} lead to the tuning system we call diatonic, in which an octave can be expressed in its simplest form as {24, 27, 30, 32, 36, 40, 45, 48}, the modern Major scale using the smallest possible numbers. The unit of 720 is given to represent a number of miles (5280 feet), and 720 is the earliest number in musical tuning theory able to host a richer octave, with notes able to express the five scales of early modal music from a common tonic note (or "*do*" in *solfage*). In figure 2.8, on page 35, the numbers of the equal perimeter model appear to have expressed, at Göbekli Tepe, the equal perimeter geometry multiplied by 720 (or factorial 6) to make 11 equal 7,920, the number of miles in the Earth's diameter, and 720 × 3 = 2160, the number of miles in the diameter of the Moon.

This cosmic version therefore shows that the Earth and Moon are part of a far-reaching geometric idea of combining the ability of 720 (to host musical harmony) and the simplest approximation to π as 22/7. It is conceivable that the Moon, as a circle on the horizon, was equated through intuition, with the circle size 3 sitting on the circle size 11 perimeter, as the Earth's horizon; that is, the 11-circle was seen as the horizon conflated with the Earth's size relative to the Moon, rising or setting on the horizon, as the 3-circle. But to guess its multiplication by 720 should give the size of the lunar diameter as 2160, in miles. This begs the question: Is the mile a cosmic unit? The whole cosmic conjecture for design of Earth and Moon must take the equal perimeter, the mile of 5280 feet and the size of the foot as 12 inches, each then belonging to this singular design. It has become clear to me that living planets probably have an ideal surface area for there to be life of the type found on Earth. This design then delivers, according to the density of the Earth and its corresponding mass, that surface area (see page 60). It seems that the mean circumference of the Earth must be 44×12^6 feet, if π is then taken as equal to

864/275.* The equal perimeter's radius of 3960 miles times 864/275 times 2 equals 24883.2 miles, which is 44 times $12 \times 12 \times 12 \times 12 \times 12 \times 12$ feet, a result first reported in my *Sacred Number and the Origins of Civilization*.

THE NUMERICAL NATURE OF MESOLITHIC ASTRONOMY

For astronomy, the southern horizon over the Mediterranean was quite different from the skies farther north, and the island of Malta may be considered ideal for observing the stars and constellations that appear in the later Greek myths. Archaeology has provided much material evidence of huts, fires, pottery, burials, and now, carbon dates and DNA evidence, but megalithic monuments are not definable without considering their astronomical purpose. The megaliths need to be studied using tools archaeology does not use—some deriving from reimagining the techniques the megalith builders had to use to understand the invariances within the number field and within the patterns of average-time astronomy. When one counts the average time periods of sky phenomena that repeat themselves, as they are seen from the Earth's surface, many interrelated patterns can be seen that are numerical.

For the megalithic astronomer, the horizon was like the shutter of a camera in that when a planet or star rises or sets, a moment in time is captured. Similarly, when an outer planet loops in the sky at nearest approach, the full Moon in proximity captures another moment, and between loops a planetary synod can be counted in whole days and whole months. For Jupiter, 27 whole lunar months is 2 synods between 3 loops, and, for Saturn, 64 whole lunar months are 5 synods between 6 loops. The lunar nodal period of 18.618 years would seem hard to count, but it is 6800 whole days. The Saros eclipse period over 18 years can be captured by counting 223 whole lunar months from an eclipse, to predict another similar-looking eclipse. In this way, counting numbers of lesser periods like the day and lunar month revealed an average-time astronomy, which had a numerical *invariance* to it: the time periods of the celestial objects mysteriously conformed to the invariance of the number field itself, perhaps by chance but in a way that also looks to be by design; that is, a compatible subset of *abstract* number relations were being expressed in the sky

---

*The constant ratio of a circle's circumference to its diameter, approximately equal to 3.14159, and in ancient times approximated by rational approximations such as 22/7, 25/8, 63/20, and 864/275, which Fibonacci discovered was most exact.

as the time intervals between celestial periods, these being counted in days or months. Celestial relationships in time appear to have naturally followed on from Stone Age interests in how numbers interact with each other within their geometrical art, strongly suggesting that the world of planetary time was a creation made using numbers. This is a perfectly reasonable hypothesis by today's standards of theory making, and indeed it should be considered unlikely to have arisen by chance.

When numbers become the size or number of physical objects, the numbers have concrete consequences, especially when the reentrant processes involved are the planetary orbits defined—we only now know—by the gravitational fields surrounding celestial objects or by their rotation or tilt. These recurring forces have generated an enduring world of eternal patterns between the planets, defined by number. The abstract patterns caused by number are due to the behavior of the cardinal numbers themselves {1, 2, 3, 4, 5, 6, 7, . . .}. The divisibility of a given number is especially important, because all numbers are made up of one or more prime numbers. For example, 6 is 2 × 3, but 7 cannot be divided, and so it is called a prime number. Prime numbers are always odd since 2 is a factor of all even numbers. So it follows that the prime numbers have spaces between them that *must* therefore be filled by second-order compound numbers—which are *not* primes. The factorization of numbers became important to the megalithic builders as a way they could work with whole numbers formed as rational fractions such as 3/2, which, when multiplied by 4/3, equals 2 (the octave). Within multiplication there is a law of conservation, that a prime number can only be removed from a compound number by dividing it with that prime number. For example, 35 = 5 × 7, and 35 ÷ 7 leaves 5, so a factor of 7 can only be removed by 7 itself.

This has led me to say that prime numbers dominated megalithic thinking, while our arithmetic plows on regardless like, Adam's fate, since our "real" numbers with a fractional part are all functionally alike with not much intelligible structure on the surface; when multiplied they are simply crunched with each other and the result seen by our decimal base of 10, whose factors are 2 times 5, give base-10 positional notation such as 17.25, which is 17 (a prime) plus 25 hundredths. As a rational fraction, 17.25 is 69/4 and 69 is 3 × 23, a prime number.

The megalithic astronomers probably lacked any bases like decimal, or positional notation as in 17.25, and so they worked with integers within counted lengths or, failing that, rational fractions such as 8/7 or 12/11, within a calculational metrology realized as, for example, Royal foot of 8/7 feet of the Egyptians and the 12/11-foot of the Sumerians. This management of numbers as integers happens to suit the synodic periodicity of the planets because they (on average) appear to circle around the observer until returning to where they started off then to repeat the same cycle.

When a cycle was measured in days, it would have been counted to make a "measurement," such as 365 (= 5 × 73) days, the Maya *haab,* or "practical," year of whole days, the synod of Venus being 8 × 73 or 584 days. Two different periods, counted in this way, are comparable using the geometry of the right triangle, whose two longest sides are the two periods. Geometrical and metrological operations can home in on their essential relationship to each other to present the two time cycles as a single rational fraction.

The Moon is obviously interlocked with the synods of Jupiter and Saturn, which are 13.5 and 12.8 lunar months respectively, in ratio as 135/128. Thirty-two lunar months are found to be 945 days, so the lunar month is 945/32 (29.53125) days long. The tropical day is also locked to the Venus synod through 73 days, which divides into both 584 days and 365 days in the ratio of 8/5, seen as the 1.6-year cycle of Venus phenomena such as evening then morning "stars" and the 8 Earth years in 5 Venus synods.

Returning to Jupiter and the lunar year of 12 lunar months, the ratio is 9/8 (the musical whole tone) in its simplest form, but in lunar months it is 13.5/12, or over two such periods 27/24, and immediately a simple whole-number ratio is findable, which will later be shown to be the highly significant musical interval 9/8, where astronomy appears musical. If one uses a foot of 9/8 feet (a unit used in the Parthenon, see fig. 6.7 on p. 127), the lunar year is 12 feet and the Jupiter synod (of 398.88 days) is 12 feet of 9/8 feet—a foot called a "pygmy" since Herakles brought small pygmy men back from India, or so the myth goes—and just larger than 1 foot. In Greek myth Herakles also performed twelve labors (the zodiac), pointing to the fact that Jupiter sweeps out, on average, 1/12 of the ecliptic in 361 days. The lunar year is also twelvefold as lunar months, while his pygmys are the enlargement by 9/8 of the lunar month, into 1/12 of his synodic period. Such is the glue inside myth pointing to a megalithic number science, because megalithic wisdom was at the root of our oldest surviving texts, oral or written.

Had the Stone Age not been able to work with integers, counting, and triangle-based or metrological ratios (the "Mesolithic package," if you will), and if these tools and methods had not been met with a high level of simple orderliness in the form of the geocentric time periods they studied, you would, for many reasons, not be reading this book, and, indeed, your history would have been quite different. This is, in part, why modern astronomy cannot believe how far megalithic astronomers were able to go accessing such a limited set data points defined when astronomical events are on the horizon or as loops among the stars. But in terms of measuring the average geocentric motions of celestial bodies, the megalithic builders were able "to reach even unto heaven." These days heaven is not so much in the sky but rather is an abstraction, such as God's heaven. Archaeology has therefore, to say the least, been blind to megalithic astronomy since it is not like the scientific astronomy we have today. This impasse perhaps explains the current glass floor between history and prehistory regarding the latter's intellectual history, only visible when their monuments are seen to be both numerate and astronomical.

RESEQUENCING PREHISTORY

While archaeologists assume the builders of megalithic structures in the west of Europe were men—that is, patrilineal pastoralists—this group had not quite arrived on the Atlantic coast, and when they did arrive these pastoralists would have been completely preoccupied with finding and farming good enough land to feed their families. It is further assumed that such monuments had a largely symbolic local meaning and could not possibly have been about finding the cosmic constants that would end up fueling later religious thinking and forming the symbols and numbers found in stories of the gods as a canon of sacred geometries and numbers.

It is impractical and implausible for the Neolithic movement of farmers to have built megaliths because it was hard, away from the Nile or Mesopotamia, to do more than subsist. It also seems impossible or highly unlikely that Neolithic farmers could have been mobilized (by whom?) for the building of large megalithic structures. Should a man wish to build a megalith, how is he going to find and recruit a group of male farmers to do some megalith building? The power structures that coalesced around Neolithic food surpluses to the east were only just becoming possible in Mesopotamia by 3000 BCE, by which time the megalithic movement was in its late period and scattered along the coasts and hinterlands of the Atlantic and Mediterranean.

In a matrilineal tribal economy, matrilineal tribespeople could field a formidable workforce where each of the women and men were closely related and competent foragers. But patrilineal farmers had moved away from foraging and so had to place subsistence above all else and make food for the whole year in one place. And if the megaliths are seen as having an astronomical purpose, facilitated by an advanced approach to numbers, then from where did the Neolithic get that motivation and facility with numbers?

The best evidence for the gender organization of the builders is perhaps the decorative and distinctive art of the goddess. Where megalithic art has survived, it is congruent with the art of the goddess in general. The megaliths, the figurines of mother goddesses, the matriarchal civilization of Crete, the Greek myths, and the early patriarchal histories and epics, all provide an obvious conclusion that the matrilineal tribes built the Mediterranean megaliths and—extrapolating from that—the Atlantic megaliths, too; and they were an intercommunicating maritime movement emanating from within the Mesolithic Old Europe. The safest early carbon dates of their monuments* are around 5000 BCE in both western France, Sardinia, and western Spain,[3] and their astronomy required an innovative mix of non-analytic geometry, foot-based metrology, non-arithmetic numeracy, and a high-performing intelligence. The culturally preferable option remains, to ignore the exact tools and techniques required or the set of astronomical facts pointed to—but unrealized by—modern astronomical data. Serious astronomy in the Stone Age has so far been unfairly ruled out as preposterous, and the patriarchal strata of modern society makes a matrilineal astronomy in the megalithic nothing less than a fully functioning taboo.

It is far more likely an elite set of matrilineal astronomers (male and female) grew a geocentric astronomy that survived into the later patriarchal world through groups like the Pythagoreans, early Christians, and the medieval geocentric planetary model called Chaldean. Later religious stories and sacred numbers were thus integrations informed by the discoveries of megalithic astronomy. These became a set of sacred symbols and archetypes, pointing into a divinely formed planetary world that appeared to be made according to numbers, seen as lengths of time, by a creator.

---

*Safest according to the methods presented in Paulsson's *Time and Stone*. She collected available carbon dates for megalithic buildings, and then filtered out samples that might have been introduced since the building was in use.

2

THE MEDITERRANEAN TRADITION

The earliest stone circle found near the Mediterranean coast is Göbekli Tepe, reliably dated to around 10,000–9,000 BCE, because of the absence of pottery in its foundations or backfill, pottery being an innovation of Neolithic-style farming. The site is therefore categorized as from the oldest period of the Pre-Pottery Neolithic, called PPN1.

Göbekli's enclosures (see fig. 2.1), could be temples, cosmograms,[*] observatories, or all three. They have a rounded style similar to the megalithic temples and observatories of Malta, and their roundness probably references the world of time. There are therefore links through Göbekli to a long-lasting Mediterranean tradition of building stone enclosures, originally for astronomical observation, but then also for memorialization of the time relationships found in the sky. Using published plans, the same 12/77-foot subunit found at Göbekli Tepe[†] can also be found in Malta[‡] at the Mnajdra observatory temple overlooking the southern sea to Africa.

---

[*]From Wikipedia, "A cosmogram depicts a cosmology in a flat geometric form. They are used for various purposes: meditational, inspirational, and to depict structure—real or imagined—of the Earth or universe. Often, cosmograms feature a circle and a square, or a circle and a cross. The circle may represent the universe, or unity, or an explanation of the universe in its totality—whether inspired by religious beliefs or scientific knowledge. The square or cross may represent the Earth, the four directions. The center may represent the individual."

[†]Schmitt (d. 2014) and Bachenheimer, who led the work, and Haklay and Gopher, who analyzed the enclosures, statistically located their centers and found the equilateral triangle linking them. See Bachenheimer, *Gobekli Tepe,* and Haklay and Gopher, "Geometry and Architectural Planning at Göbekli Tepe."

[‡]See D. H. Trump, 2002. Trump was a long-term archaeologist in Malta who worked with Evans in Crete. In 2014, Tore Lomsadalen visited Malta to establish sightline data to the Sun and Moon, but not to the bright stars. See his *Sky and Purpose in Prehistoric Malta.*

FIGURE 2.1. Enclosures B, C, and D of Göbekli Tepe. Enclosure C was shaped by three overlapping circles to form a rounded rectangle. Photo: Göbekli Tepe Project.

The 12/77-foot subunit enables an innovative numerical fact useful when constructing circular structures, and it is a primary unit found in use in 600 CE by Cappadocian monks (see chapter 7), who had churches carved out of rock and which followed architectural plans that were diagrams of the most significant cosmic time cycles.

Comparing the Göbekli enclosures and Cappadocian churches by reason of a shared unit, one is faced by a double anachronism of a seemingly megalithic Göbekli Tepe built long *before* the megalithic period and early Christian monuments built long *after* the megalithic period. Yet they all used a special subset of metrology based on the 12/77-foot subunit.

FIGURE 2.2. Mnajdra in Malta. Compare with Göbekli Tepe in figure 2.1.
The enclosures are both based on multiple circles with diameters based on
the 12/77-foot subunit.

No megaliths older than around 5000 BCE have been found in western
Europe, so what does Göbekli Tepe represent? Initially thought to be megalithic
stone circles, its enclosures were contemporaneous with a cluster of similar
settlements on the northern uplands, exactly where the Neolithic package was
starting to be developed. And it was Neolithic farming that enabled the historic
civilizations of the ancient Near East to develop to the south, in Mesopotamia,
after 3000 BCE. By 5000 BCE, the Neolithic package had also migrated from
the Near East to Central Europe and then the Paris basin, but not to Carnac;
that is, the geographical appearance of the megalithic package, so to speak,
of monuments and number sciences appeared on Europe's coastal hinterlands
around that same time of 5000 BCE.

The rounded form of the enclosures have rings of stone pillars that are regularly placed and two T-shaped stones at their center. Their oval shape, as at Mnajdra, derives from multiple circles and the number of enclosures is organized as an equilateral triangle, but there the enclosures appear organized for astronomy to the south while the Göbekli enclosures, as a triangle, seem to point north. We will interpret this later as indicating that Göbekli was a manifestation of polar astronomy rather than of the subtropical astronomy of Malta.

INITIAL ANALYSIS OF GÖBEKLI TEPE

An analysis of enclosure C revealed the presence of the likely and very useful architectural subunit of length 12/77 feet, an innovative unit of measurement that divides into at least three of the most prominent foot-based measures of the later ancient Near East,* as follows:

7 of these units form the Sumerian foot of 12/11 feet;
11 of these units form the Royal cubit of 12/7 feet; and
44/7 ($2 \times \pi$) of these units form the common Egyptian foot of 48/49 feet.

We know that the English foot became the standard foot of a single, "ancient" metrological system. Indeed, all historical metrologies can be derived as having had an astronomical genesis when counting days as inches; this was clearly present within the Le Manio Quadrilateral at Carnac, in Brittany.[1] However, if 12/77 feet was in use at Göbekli, the foot-based metrological system must already have been in use prior to its creation—a fact upon which the overall form of prehistory might be deduced (see chapter 8). If Göbekli Tepe was employing this metrology before both—the megalithic period after 5000 BCE, or the Neolithic civilizations after 3000 BCE— then Le Manio must have been a recapitulation of a remembered astronomical derivation, which means that foot-metrology existed before the end of the Ice Age.

---

*One should note that a worldwide metrological system existed (see John Michell, *Ancient Metrology,* and John Neal, *All Done with Mirrors*), in which every measure is a rational fraction of the English foot. It appears to have been used for megalithic-style astronomy. In this system the English foot is the root foot of a Greek module of feet in all later historical metrology, but the English foot was also the root of all the other modular roots, these being simpler fractional variations of the English foot, through which a proportional numeracy was achieved.

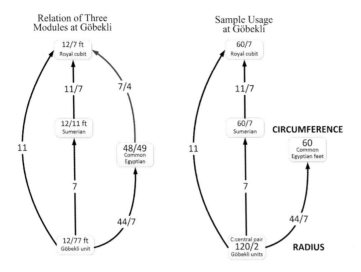

FIGURE 2.3. *Left,* the relations of the Göbekli subunit to three ancient measures; and *right,* the relation of a circle's perimeter to the radius creating it, using the 12/77-foot subunit.

Laying anachronistic mysteries to one side for the moment, we find supporting evidence for the efficacy of this subunit in use at Göbekli Tepe, allowing its builders to numerically relate the perimeters of circular structures to their radius (or diameter) without any arithmetic, because the ratio 12/77 embodies the two prime numbers ($7 \times 11 = 77$) often used by the megalithic as an accurate approximation to $2 \times \pi$; namely, 44/7.

Given the basic state of human numeracy that is presumed in Pre-Pottery times, this fortunate subunit of 12/77 feet would give the length of a circle's circumference without any calculation. Circles could be drawn using a staked rope, and the number of 12/77-foot subunits in the radius rope would create the same number of common Egyptian feet (of 48/49), as in the circle's perimeter (or circumference). Radii in these subunits were therefore a very effective metrology, and basic numeracy would have sufficed, using such direct shortcuts instead of the more complex methods of later Near Eastern metrology (see fig. 2.3 *right*).

For example, from an available plan of Göbekli's enclosure C (see fig. 2.4), there are 120 subunits between the two central stones so that a circle of that diameter would have a circumference half of that, of 60 common Egyptian feet in length (58.7755 ordinary feet). Having counted out 60 subunits from the proposed center in between, a circle drawn manually (using a rope staked to the center) could be known as "sixty" by both radius and circumference. We shall see later that these midpoints between the twin pillars of enclosures A, C, and D were formed to be the points of an equilateral triangle, so that the radii from these would probably have defined the placement of the pillars.

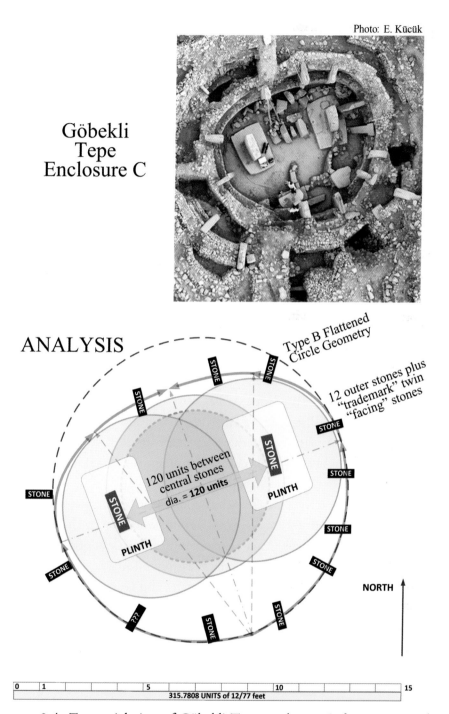

Photo: E. Küçük

Göbekli Tepe Enclosure C

ANALYSIS

Type B Flattened Circle Geometry

12 outer stones plus "trademark" twin "facing" stones

120 units between central stones
dia. = **120 units**

STONE

PLINTH

PLINTH

NORTH

0   1   5   10   15
315.7808 UNITS of 12/77 feet

FIGURE 2.4. *Top,* aerial view of Göbekli Tepe, enclosure C; *bottom,* an analysis shows its triple circles and the 120 subunits between its two central pillars. From Haklay and Gopher, "Geometry and Architectural Planning at Göbekli Tepe," 349, fig. 4. For further analysis, see the section "Interpreting Göbekli Tepe" on page 33.

A MATRIARCHAL MEGALITHIC

Similar building practices at Göbekli Tepe and the Maltese enclosures might connect the two sites culturally, despite their location at opposite sides of the Mediterranean's eastern basin and their separation of many millennia in time.

The circular geometry of the enclosures of Malta perhaps followed the lines of the rotund body of the matriarchal goddess, seen in the classic Ice Age veneration of voluptuous Venus figurines that would seem to celebrate the eternal role of women as genetrices, child-rearers, and horticulturalists. And astronomical time is endlessly being created. The circular walls of the stone cairns of Göbekli Tepe and their use of sculptured stone lintels are similar to the form of the linteled walls of Malta's temples. Mnajdra in particular was an observatory with alignments to horizon events for Sun, Moon, and stars over the southern Mediterranean. Holding vital clues to the unique but underappreciated role of women in the megalithic, Malta has instead been presented by disconnected historical narratives, downplaying the female role.* As stated in the introduction, because of bias the Maltese builders have been thought to be Neolithic rather than Mesolithic. In the Mesolithic, tribes were normally matrilineal foragers of fauna and flora. As discussed below, it may seem more likely that Göbekli Tepe's builders were foragers but perhaps not matrilineal. But, as discussed earlier, matrilineage was a survival or efficient strategy, yet seeing it in the past is always a last resort for a culture steeped in male patronage.

The use of goddess figurines in Malta confirms that the megalithic enclosures of Malta were built by a matrilineal culture whose many-themed art was distinctive across Old Europe,† art that combined female motifs with purely geometrical patterns, some of them perhaps astronomical.

The concrete link between megalithic Malta and Göbekli Tepe is the unit of 12/77 feet in use for the circular form of the Malta monuments and the more elliptical triple circles of Göbekli's enclosures C and D. The underlying astronomical symbolism at Göbekli, as well as the equilateral triangle,‡ shows a type of knowledge shared with later megalithic sites, knowledge natural to

---

*See Rountree, "The Case of the Missing Goddess: Plurality, Power, and Prejudice in Reconstructions of Malta's Neolithic Past." She documents some of the local and academic obstacles to recognizing the obvious goddess culture behind Malta's megalithic structures.

†See Marija Gambutas, *The Language of the Goddess.*

‡An architectural grid that links the three enclosures.

average-time horizon astronomy wherever and whenever it was practiced. Göbekli Tepe's enclosures seem like meeting places with troglodyte benches that, enable viewing the sky in all four directions through the punctuation of tall flat slabs aligned to sightlines out of the circle.

THE SLEEPING GODDESS OF MALTA

Using circles around a backsight, alignments to the sky are suited to establishing radial alignments to the horizon events involving the Sun and Moon and the study of the whole night sky above the "plain of the sea." Insight into the emergence of stellar astronomy in megalithic Malta can be seen in the form of the "sleeping goddess" (see fig. 2.5 on p. 30), a statue found in Malta's unique Hypogeum, a rock-cut cave that was almost certainly a unique and faithful subterranean replica of an aboveground temple, thereby preserving the lost form of Maltese temple design, now ruined above ground.*

The meaning of this original sleeping beauty correlates with the Olympian creation story:

> At the beginning of all things Mother Earth emerged from Chaos and bore her son Uranus as she slept. Gazing down fondly at her from the mountains, he showered fertile rain upon her secret clefts, and she bore grass, flowers, and trees, with the beasts and birds proper to each. This same rain made the rivers flow and filled the hollow places with water, so that lakes and seas came into being. (Graves 1955)

As has been found to be the case with many myths, this story may have pointed to something concrete—namely, Malta's role in innovating stellar astronomy, before the composition of the Greek myths as we now know them. The first description of the constellations was by Eudoxus, allegedly from an antique star globe given to him in Egypt. The scope of the constellations he described, and those mentioned in the Greek myths, belong to an original set of observations made at 36 degrees north, the latitude of Malta around 2800 BCE. Only the set of stars visible above the South Pole at that date were shown on that globe, which inspired the first star charts. A modern review proposes this

---

*Other unique models and relief carvings around the temples of Malta have further enabled missing upper details of the temples to be visualized.

FIGURE 2.5. Museum reproduction of a recumbent goddess, whose voluptuousness follows the line of the celestial equator and ecliptic mountain of heaven above the plain of the Mediterranean, through her hips. Note how her dress has areas resembling the modern constellation boundaries, where objects are catalogued as belonging to the constellation they fall inside of.
Photo: A. G. E. Blake.

astronomical work as having been done for "navigators," and this date is far too early to be the Phoenicians, who were first-millennium maritime traders. The origin of the constellations may have been associated with the sea routes linking megalithic regions, and, if so, those navigators were also astronomers. The constellations calibrate the rotation of the Earth and the notion of latitude and longitude.

If the goddess is the sky, her central axis is the celestial equator and her rising hips symbolize the rise and fall of planets along the Sun's path, or ecliptic, which the zodiacal stars* punctuate. Such a view would mean looking south. The flat table upon which the goddess is sleeping is the Mediterranean Sea, perhaps from Mnadjra's observatory. Her head is then to the west and her feet to the east.

---

*The zodiacal constellations were not from Greece. Early, southeastern Mesopotamia developed these within an agricultural calendar of helical risings, constellations rising before the sun.

STELLAR ASTRONOMY AT MALTA

Mnajdra appears to have been built in a similar fashion as the older monument of Skorba, farther north,* but with the added feature that the horizon was seen from near sea level (see fig. 2.6, *top*). The alignment of the four-stone entrance way (similar to Mnadjra's, see fig. 2.2) is toward Rigel on rising from the sea, the bright star of Orion's right (or western) foot. The axis of the middle enclosure follows the eastern jambs of the central passage, which are aligned to the rising of Sirius above the sea. Figure 2.6 on page 32 shows a star map from a planetarium program, showing the southerly view in which the star called Saiph (Orion's left foot) can be seen to form a southern meridian, vital to the making of star maps as the stars move west, up to the eastern belt star of Orion.

The view south from Malta was surely most amenable to any megalithic culture when engaging in the task of viewing the path upon which the Sun, Moon, and planets all slowly "wander" from *west to east* on the ecliptic. The celestial equator defines the parallel direction of star movement from *east to west* every night; the "fixed" stars move together as a single pattern, as if fixed to a black sphere beyond them. These constellations become easily recognizable and are given human or animal names, above and below the celestial equator. The bright stars are useful landmarks in this celestial reflection of the Earth, especially those near to the Sun's path, which then divide the ecliptic unevenly as does our zodiac of twelve equal signs. Looking south over the sea, both the form of the star constellations and how far the stars were from each other in angle could be quantified using a naked-eye meridian sightline from Mnajdra's northern enclosure, to the east of the other two enclosures.

In Indian astronomy the Moon visits 27 or 28 bright-star *nakshatras* (lunar mansions) in the course of her orbit. In fact, the Moon's orbit is 27⅓ days so that three orbits take 82 days—enabling the orbit to provide an opening to the greater task of quantifying the relative positions of the stars on the fixed sphere. That work had to have been started at this latitude.

To count the duration of a lunar orbit in days (27⅓ days), one needs to see any prominent stars that the Moon is passing by in her orbit. This sort of procedure is implied in Indian astronomy, and other traditions, where the sky was first divided into 28 or 27 equal parts, called lunar mansions (nakshatras).

---

*As surveyed by D. H. Trump and Sir Arthur Evans, who discovered the Minoan site of Knossos in Crete. See Trump's *Skorba: Excavations Carried Out on Behalf of the National Museum of Malta.*

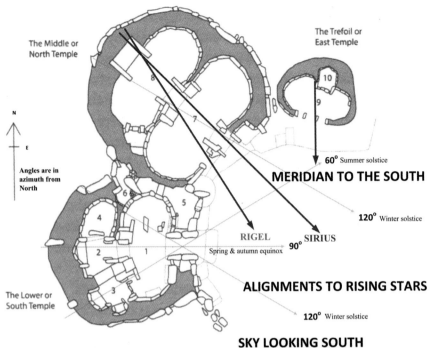

FIGURE 2.6. *Top,* the site plan of Mnajdra showing that the older eastern trefoil resolved the rising of Sirius and Rigel on the sea's southern horizon. *Bottom,* looking south with the eastern leg of Orion marking the southern meridian at the rising of the star Saiph, between Sirius and Rigel in declination.

It was also easy to see the loops of the planet Saturn every 378 days, and there are just less than 28 loops in 29 practical years (365 days), which means the daily lunar movement is nearly equal to the movement of Saturn's loops, so that Saturn has a number association with the Moon. Only the Indian star chart has retained the lunar mansions as we see in figure 2.7 on page 34.

Yet again, one sees that many things were possible to a megalithic astronomy that may have resulted in their unnoticed transposition into later aspects of the historical record due to our ignorance of those achievements.

INTERPRETING GÖBEKLI TEPE

Using the published sources, the 12/77-foot unit appears to calibrate Göbekli Tepe, in particular the scaled plan of enclosure C.[2] The three enclosures B, C, and D all have twin dressed-stone slabs at their centers and, when these were measured off the plan for C, the distance between their inside-facing surfaces appeared to be 120 units of 12/77 feet. This would give the earliest use of this previously unknown unit (see fig. 2.4). It is also true that Göbekli Tepe, being at least 8500 years old, could even be older—and from 9880 BCE—given its alignment of 15 degrees north (see the section "Göbekli Tepe's Northern Axis of 15 Degrees" on page 42). It is therefore the oldest monument considered here. To perform its circle magic the unit must be 12/77 feet long, but beware: the unit 12/88 feet is also key to the overall geometry of the three enclosures and the equilateral triangle found between them.

The Equal-Perimeter Model
As already stated, the equal-perimeter geometrical model[3] (see figs. 1.3 and 1.4) is based on π as 22/7, and, in its lowest numeric form, the two concentric circles have diameters 11 and 14. The inner circle is the mean Earth (11) diameter and the small circle (3) make a pair of circles that mimic the 11 to 3 ratio between the diameters of the Earth and the Moon. To obtain an absolute scaling in units of miles, these numbers must be multiplied by 720, which is 6!, or $1 \times 2 \times 3 \times 4 \times 5 \times 6$ terminated by 7, the first of the two "circular" prime numbers, 7, and 11, found as their product in the denominator of 12/77; that is, 77 as 7×11.

The factorial form of the equal-perimeter geometry appears at Göbekli Tepe by subtle inference in the heights of enclosure D's inner ring of "exactly twelve" identical T-pillars, which are all 4 meters high. One can see this height as 12 Sumerian feet of 12/11 feet, equaling 3.99 meters. Seen as rational factors,

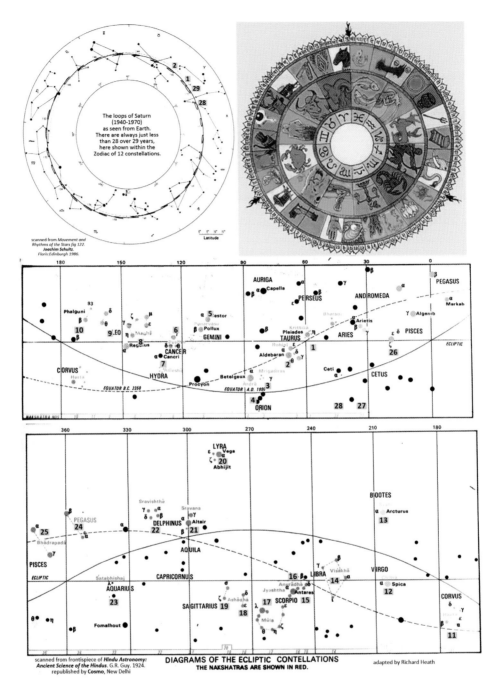

FIGURE 2.7. The type of astronomy suited to Malta would involve a chart or set of distances between bright stars, allowing for the study of planetary loops and also the foundation needed for the star constellations. *Below,* is the Indian star chart of the nakshatras. *Top right* is a wheel of nakshatra symbols, with the twelve signs of the zodiac within it. *Top left* shows the 28 loops of Saturn, which resemble the daily motion in the sky, of the Moon in her orbit. The lunar mansion on a given day was identified by the name of the lunar mansion seen behind it. In 2350 BCE, the system started with the spring equinox, which stood on the seven sisters or Krittikas.

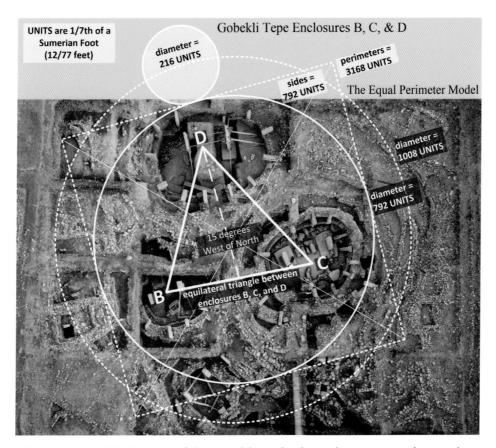

FIGURE 2.8. Likely sizes of the possible multiple circle structures that might have been used to locate enclosures B, C, and D, couched in the equal-perimeter model's square and circle. Enclosure B appears to represent the Moon through its diameter of 216 units. Aerial photo: Haklay and Gopher, "Geometry and Architectural Planning at Göbekli Tepe."

12 stones × 12 Sumerian feet × 12/11 feet equals 1728/11 feet, or 1008 subunits of 12/77 feet. And 1008 is the smallest integer diameter possible for the equal-perimeter circle (of length 14) in whatever units—and in decimal notation—and its factorial version of 720 (equaling factorial 6) has been reduced by dividing by 10, to give 1008. The outer circumference is then 3168 subunits, or a radius 504 leading to 504 common Egyptian feet on the circumference—this circumference being known from the "number" of the radius rather than deduced through any calculation, no further measurement being required. The diameter 1008, divided by 11/14, gives an inner 11-circle of 792 subunits, the equal perimeter of its out-square then being 4 × 792 or 3168 units. This

number, as a perimeter, is that found in the Neolithic and later periods onward as the proper perimeter of a temple or sacred space, which in turn represents the sublunary sphere of the geocentric model. The inner 11-circle of 792 units is the mean Earth diameter in miles, and the 3-circle of 216 is the diameter of the Moon, all in units of 10 miles per unit. This Göbekli version is therefore a 72-times table of the basic dimensions, ignoring the zero in factorial 6 as 720.

This interpretation finds the builders of Göbekli Tepe using the metrology of the English foot five-thousand years before the megalithic buildings of Carnac, where it will next be seen, in this book. The factorial-sized geometry of fig. 2.8 models the Earth and Moon in 10-mile units by upscaling the basic form 11/14 by 72 units, a kind of initiatory revelation likening π to the presumed cosmology of an inspired tradition, numerically maintainable by small pre-arithmetic tribes belonging to a Mesolithic culture and transmitted as an eminently portable form of knowledge. The available aerial photos may have distortions, but work on the three enclosures using it, based on the 120 units between the central pillars of enclosure C, resulted in an interesting set of tentative quantified geometries from fig. 2.4.

Enclosures C and D are not circular, inviting the use of triple overlapping circles to form their unique shapes. But enclosure B seems intentionally circular and has the size (216 units) associated with the Moon's diameter of 2160 miles.

This fact—that the three enclosures were so well framed by the cosmological equal-perimeter model and using the lowest possible head numbers*— seemed promising to my endeavor, since, as John Michell observed, this model was often used in the ancient world as a sacred boundary for representing the geocentric world as a geographical center. These three enclosures might, by good fortune, be the sacred space within Göbekli Tepe.

The Equilateral Triangle between Enclosures

As previously mentioned, the three enclosures are precisely located with centers forming an equilateral triangle whose east-west base is about 15 degrees north of east, implying alignment to a circumpolar star (Vega), to be discussed in chapter 8. The foot ratio 12/77 appears as the main Göbekli architectural unit, and in the first millennium these units are again seen in chapter 7, giving additional

---

*Head numbers are viewed separately from a tail of noughts in decimal integers since they contain numbers that have effectively been shunted by the noughts, through scaling by powers of 10. They often correspond to the sacred numbers found within texts such as 432,000, where 432 can support a pentatonic tuning system, or the root Yuga unit, which then doubles to 864, 1728, and so forth.

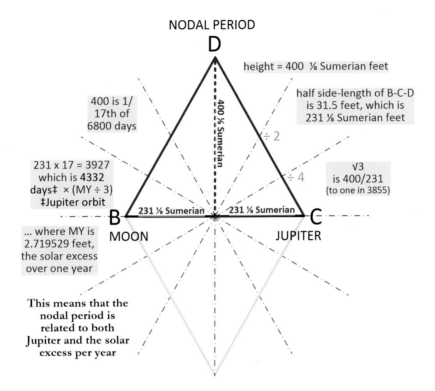

NODAL PERIOD

D

height = 400 ⅛ Sumerian feet

half side-length of B-C-D
is 31.5 feet, which is
231 ⅛ Sumerian feet

400 is 1/
17th of
6800 days

400 ⅛ Sumerian

÷ 2

231 x 17 = 3927
which is 4332
days‡ × (MY ÷ 3)
‡Jupiter orbit

÷ 4

√3
is 400/231
(to one in 3855)

231 ⅛ Sumerian   231 ⅛ Sumerian

B

C

... where MY is
2.719529 feet,
the solar excess
over one year

MOON

JUPITER

This means that the
nodal period is
related to both
Jupiter and the solar
excess per year

Equilateral Triangle between enclosures B, C and D of
Gobekli Tepe as 1/17 model, of the nodal cycle relative to
Jupiter's orbit of (4332 days), in solar days.

FIGURE 2.9. Haklay and Gopher's parametric centers of enclosures B, C, and
D were found to form the vertices of an accurate equilateral triangle whose
axis is aligned to 15 degrees west of north. From "Geometry and Architectural
Planning at Göbekli Tepe." MY is over THREE years

extensions of the equal-perimeter model in some of the rock-cut churches of
Cappadocia in central Anatolia.

Using state-of-the-art computer analysis, Haklay and Gopher found the
likely centers of enclosures B, C, and D and found these centers connecting the
enclosures via a near perfect equilateral triangle (see fig. 2.9).[4] A correspondent*
brought to my attention a foot-based metrology within Göbekli Tepe. Before
this my inclination was to assume that the Carnac megalithic culture had
first defined metrology around 5000 BCE. This would open the door to an

---

*Fred Morris Jr.

"Atlantis," a culture prior to Göbekli Tepe that could be the fabled Arctic center, possibly during the interglacial period, to be discussed in chapter 8.

The equilateral triangle* has sides of length 63 feet or 404 units.[†] Any equilateral triangle can be divided into two equal {1, 2, √3} triangles about its vertical height, making the half side 31.5 feet, which multiplied by √3, is 350 units. But if another unit had been defined, one-eighth of the Sumerian foot (12/11 feet) equaling 12/88 feet (this then reduces to 1.$\overline{63}$ inches) The height would then be 400 units of 12/88 feet because the new unit is seven-eighths smaller, making 350 seven-eighths larger, as 400 units. Because both units divide the Sumerian foot, I suggest the new unit can be noted as SF/8 in contrast to the first subunit of SF/7. Applying SF/8 to 63 feet yields 462 subunits. One sees an interesting approach, not so far seen in ancient metrology, in which the foot measure used to make a measurement can be adventurously varied to form suitable subunits to manipulate the prime factors within the measurements. One normally thinks of a foot as having a singular unit, such as 12 inches or 16 digits. After all, the units, subunits, and aggregation of measures (such as yards) employed were being applied to make astronomical counts rational with respect to each other, as a pattern.

It now seems likely the megalithic astronomers of 5000 BCE inherited at least a partial system of foot-based measures from a source related to Göbekli Tepe built five thousand years before. The simplest explanation would be that the Mesolithic peoples of Europe, based on an oral transmission from before the terminal Ice Age maximum, may have placed more emphasis upon fewer units by using subunits like SF/7 and SF8 within their standard geometries, to reduce complexity.[‡]

In this case, if the height of the equilateral triangle is 400 SF/8, 400 divides into the 6800 days of the Moon's nodal cycle seventeen (17) times, and if 231⅛ subunits are multiplied by 17 then 3927 is the result. This 3927, divided by the

---

*In megalithic Wales, I had only recently found an equilateral triangle relating, through measures, to the Moon's nodal period (of 6800 days) to a product of Jupiter's 361-day year (= 19²) times the 10.875-day excess (= 87/8) of the solar year over the lunar year in one solar year (see chapter 3). Thence came my initial interest in Göbekli Tepe's equilateral spacing. The nodal period is therefore related to the orbital period of Jupiter, which is 4332 days.

†The number 404.25 is very close to 63 × 12/77 feet, which is 404.

‡The comparison of time periods as ratios used the geometry of triangles, the right and equilateral triangles related to the trigonometry found between points on the circumference of circles relative to cardinal directions at right angles. The mathematical character of triangles (called *trigon* in Greek) was only implicit until the third century BCE, when the Greeks discovered the *trigon*-ometry of sines, cosines, and tangent ratios. The megalithic use of triangles was, however, not mathematical as such, but was established through discovered comparisons and coincidences.

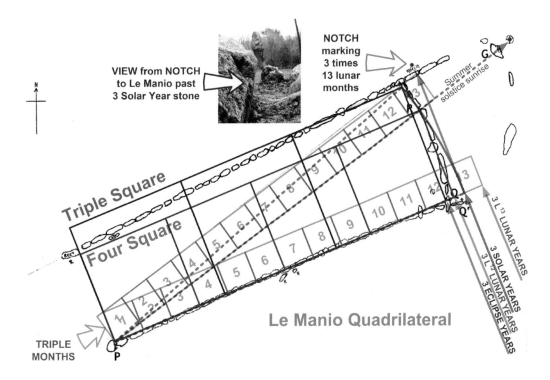

FIGURE 2.10. Picture of Le Manio near Carnac. This megalithic yard was generated at Le Manio as the excess (in day-inches) of the solar year over the lunar. For this measurement a {9, 36, 37.1} lunation triangle was employed, perhaps seen as half of a 4-square. See Richard Heath, *Sacred Number and the Lords of Time,* figure 3.8.

Jupiter orbital period of 4332 days, gives the result 0.906509 feet, which is one-third of the megalithic yard of 2.719529 feet.*

Could Göbekli Tepe's three enclosures B, C, and D represent the Moon's nodal relationship to Jupiter, as modified by the excess of the solar month (30.437 days) to the lunar month (of 29.53059 days) over a single mean solar month?

To understand the genius of equilateral placement of these enclosures, one must look at the ratio 3927/4332, which accurately equals 29/32, since both 29 and 32 figure strongly in the geocentric pattern of time.

---

*The same result was found in the Preseli Hills in western Wales, using a 1-2,-√3 triangle north of Pentre Ifan (see chapter 3, and fig. 3.3 on p. 51), only in that case the geometries were massive, being based on English yards.

Dimensionality of Göbekli Tepe

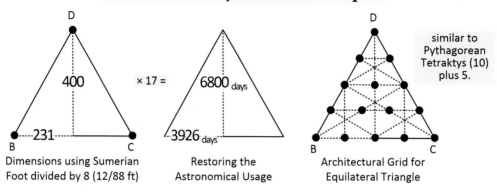

Dimensions using Sumerian
Foot divided by 8 (12/88 ft)

Restoring the
Astronomical Usage

Architectural Grid for
Equilateral Triangle

similar to
Pythagorean
Tetraktys (10)
plus 5.

Enclosures {B C D} form an Equilateral Triangle of centers,
where height over half base is 400/231 = √3

FIGURE 2.11. *Left,* the three enclosures in units of Sumerian foot divided by
88, called SF/88; *center,* the expansion to 17 ties that as numbers of days, where
the height is then 6800 days; *right,* the likely way in which a grid of equilateral
shape could have been used and upon which the enclosures could have been
built around grid points.

The number twelve has the property of integrating the numbers three and
four, and this seems to be exploited in the three astronomical twelves:

1. The lunar year has 12 lunar months.
2. The solar year has 12 mean solar months or MSM.
3. The orbit of Jupiter, of 4332 days, has 12 Brihasparti years of 361 (19^2) days.

As to the last item, this third division of the largest planet's orbit appears
implicated in the causation of the nodal period of 6800 days through the geometry of the equilateral triangle, which is a completely new and unexpected discovery although, the fact is that the lunar year is already seen to have whole
tone synchronism with the synod of Jupiter. So, while Jupiter is interacting
synodically upon the Moon's orbit, the major gravitational influences are both
Jupiter and the Sun and we see within the Göbekli equilateral that the encoded
height of 6800 days is √3 × 3926 days and 3929 days ÷ 361 days = 10.88, the
difference between the solar and lunar years in days.

Jupiter traverses the whole zodiac in 4,332 days and therefore crosses 1/12 of the zodiac in 361 (19²) days, which is why the twelvefold zodiac became emblematic of Jupiter/Zeus within Indo-European tribes who, unlike matrilineal Greeks, had the letter Z of Zeus in their alphabet. No wonder, then, that the new god Zeus was a patriarchal name. If 4,332 days × √3 had been 6800, then this would probably be seen as Jupiter and the Sun *causing* the precession of the Moon's nodes. But 4,332 days must be multiplied by 29/32 and then by √3 (1.732), to equal the 6800 days of the nodal period. The ratio 29/32 is key to the time pattern of the Sun and Moon. As stated above, 29/32 days is the difference between the lunar and mean solar month. In 945 days, exactly 32 lunar months occur so that the megalithic astronomers could model the lunar month as a rational fraction of 945/32 = 29.53125 days, which, plus 29/32 days, equals 30.4375* days, which, in turn, multiplied by 12, gives the solar year as 365.25 days.

Why should the √3 relationship apply this factor 29/32 to the Jupiter orbital period? The influence of Jupiter's 4,332-day period seems to be reduced by the solar-lunar relationship in a shared effect upon the lunar nodes and their precession over 6800 days. The √3 relationship is 3927 using (as one must) the same unit of days, as 17 × 231 (= 3927) units of half the equilateral's side length. If 29/32 is the excess in days of the solar month to the lunar month, then this solar excess must represent the precessional effect of the Sun upon the Moon, then combined with the effect of Jupiter's orbit. The geometrical symbol is previous to the situation with the gravitational forces and therefore implies that reality is originally based on numerical criteria that must then be achieved by the celestial bodies involved; that is, reality has a supracausal origin prior to the forces of causation.

Origins of the Equilateral Model

Could the twin facing stones at the center hold some clues? In all three enclosures, these are 18 Sumerian feet (5.5 meters) high (3/2 the height of the 12 stones of enclosure D.) The stones are 144 SF8 in height. Just as the 12 pillar heights of enclosure D sum to give the 14-circle diameter for the equal-perimeter model, one needs to look carefully at how any ancient group could have come to this equilateral model as perfectly expressing how the lunar nodes are synchronous with both Jupiter and the solar excess. One standout result is that 31.5 (half 63) feet are 378 inches, numerically the length of the Saturn synod in days, and this implies an even larger matrix of underlying relations existing between the gravitational influences called planets, when seen from the Earth.

---

*1461 ÷ 4 = 365.25.

We will see in chapter 4 that the internal walls of an old basement chamber at Knossos, called the monolith room, was a 4-by-3 rectangle with two facing pillars in the middle. The unit squares then have a side length of 27 SF/7 so that the perimeter of the rectangle is 378 SF/7.

One should also note that 31.5 feet is 1/24 of the 756-foot southern side of the Great Pyramid, indicating how monuments resonate over millennia with the model here turned into 231 SF/8.

With regard to construction, the day-inch count of 378 day-inches (the Saturn synod) can, from *B* to *M*, be doubled to set up the base of the equilateral *BC*, whereupon two arcs of the same length, from either end, cross to place the third point, *D*. At that point, the midpoint of the base *BC* forms a symmetrical axis whose height is 400 SF/8 divided by *BM,* which gives √3. The axis *MD,* of 400 SF/8, is 17 times less than the nodal period of 6800 days, while *BM,* as 231 SF/8, times 17 is 3927, the combined effect of Jupiter and solar excess upon the precession of the Moon's nodal period.

Göbekli Tepe's Northern Axis of 15 Degrees

Many megalithic monuments appear to point directly, or somewhere near, north. One such is the equilateral triangle linking Göbekli Tepe's three enclosures B, C, and D, whose northerly axis points 15 degrees west of north to the circumpolar star Vega in the constellation Lyra.

In my *Sacred Number and the Lords of Time* (page 87), I was able to see northerly alignments as defining a type of astronomy rarely credited to the megalithic, which found alignments to circumpolar stars that rotate around true north. The pole rarely has a star within degrees of it, so our present pole-star, Polaris, is a rare situation. The circumpolar stars rotate once every sidereal day owing to the rotation of the Earth, and the solar day (our 24 hours) is slightly longer than the sidereal day by the ratio 366/365, a difference of nearly 4 minutes, and, over a solar year, all those excesses add up to an extra sidereal day of 366 to 365 solar days.

Sometimes, a strong circumpolar star can be used to estimate sidereal time inside the span of a day, something that all mechanical clocks do with their hands and a circular face, the hands of clocks aping the circumpolar region as the Earth rotates and Vega was the hour hand before 9900 BCE. Looking north from Göbekli Tepe, Vega would travel counterclockwise, describing a circle in the sky, and at the center was an empty pole. Vega's maximum elongation from north to the west was aligned to the center line of Göbekli Tepe's enclosures, which were aligned 15 degrees west of north. The star's name from antiquity meant "the falling one," an eagle or vulture that falls upon its prey. In the constellation Lyra, a lyre is sometimes shown as in the ÿed ÿuree's claws or beak. It is the second brightest white star, and Vega became the polestar in 12,500 BCE. By 9880 BCE it had the 15-degree range of alignment west and east of true north, and this might help date the monument within the 10,000 BCE to 8000 BCE range (early Pre-Pottery) given by archaeology, to around 9880 BCE (see fig. 2.12).

In 12,000 BCE, before nearest approach to the pole, Vega would have been "falling" upon the North Pole before moving away from it, and this might have been portrayed by the head removed from a body by a vulture, depicted in both a wall painting in Çatalhüyük and on a pillar at Göbekli Tepe, both these then showing "news" from the Ice Age "Gazette" (an oral and graphic publication) (see fig. 2.13 on p. 44).

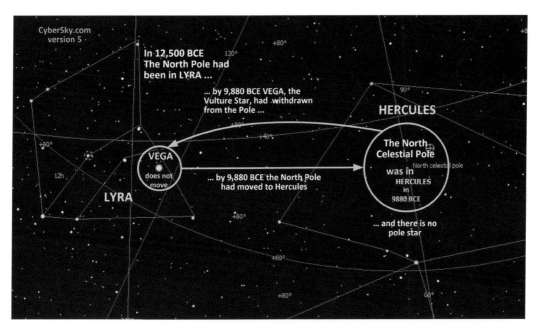

FIGURE 2.12. The circumpolar sky from Göbekli Tepe in 9880 BCE, when ex-polestar Vega would stand above northern enclosure D, viewed from the midpoint between enclosures B and C, once per day, then 15 degrees from the North Pole. The North Pole had moved owing to the precession of the equinoxes (which takes more than 25,920 years to complete), and now, no longer the polestar, Vega orbited the pole at a separation of 15 degrees. By then the North Pole was in Hercules, who must perform his twelve near-impossible labors, giving him a zodiacal association, but here at the North Pole.

If true, the polar alignment of Anti-Taurus implies the builders of Göbekli Tepe used the enclosures, in part, as a circumpolar observatory after their arrival through the Anti-Taurus mountains from a more northerly location, such as the Caucasus region or even the coast of the Arctic Sea inside the Arctic Circle. Chapter 8 discusses the possibility that during the Ice Age, the Gulf Stream kept the Arctic Sea's coasts unfrozen and habitable, in contrast with the interior of Europe, which was covered by massive ice sheets. On the Arctic shores of Europe, all the stars within the Northern Hemisphere are circumpolar, rotating "westward" above the horizon along with the northern part of the ecliptic, which latter is then visible almost all the way around the horizon during a dusk, a night and dawn lasting half a year. Such considerations have been thought to have provoked very early astronomy, like that at Göpekli Tepe, this as recorded in the Ṛg Veda (see chapter 8).

Figure 2.13. The Vulture Stone of Enclosure D is symbolic of the polestar Vega in 12,000 BCE, at the zenith to someone at the pole itself. The same vulture motif is also seen on Çatalhüyuk's walls and other related monuments, in stone relief. One might even visualize a lyre's strings in the "feathers" or strange inner organs.

3

REUNION OF THE
MEGALITHIC LATITUDES

Seven thousand years after Göbekli Tepe, phase 3 of the world-famous Stonehenge was built in what is now Wiltshire, England. There was already a large stone circle with 56 rude bluestones built around 3000 BCE, which marked the arrival of megalithic influences from Brittany, and Ireland via Wales. In the preceding millennium, the local cultures had built large straight "roads" (now called a cursus), lightly delineated by banks or ditches and punctuated by circular henges and burial barrows. The 56-stone circle was moved from Wales to Wiltshire and is now called Stonehenge 1, and this marked the transition from earthworks to megalithic structures; namely, standing stones, stone rows, and circles and dolmen. But when Stonehenge 3 was built (circa. 2650 BCE) its design was discontinuous from preceding stone circles and would remain unique in its design and intensity, using stone that fit together with jointing techniques, trilithon gateways, and stones that were sculpted. These features resemble those found in structures built in the Mediterranean and Egypt. Before carbon dating, Stonehenge in particular was thought to have diffused from the Near East, but carbon dates place it at the very beginning of dynastic Egypt. But there are some good reasons to consider Stonehenge the result of a direct visitation, celebration and reunion of the Mediterranean and Atlantic branches of megalithism.

Before Stonehenge, the "cursus," or earthworks, culture was native to Britain.*

---

*"Cursuses are monumental Neolithic structures resembling ditches or trenches in the islands of Great Britain and Ireland. Relics found within them indicate that they were built between 3400 and 3000 BCE, making them among the oldest monumental structures on the islands. . . . The Stonehenge Cursus is a notable example within sight of the more famous Stonehenge stone circle. Other examples are the four cursuses at Rudston in Yorkshire, at Fornham All Saints

A uniquely good example being Thornborough Henge in North Yorkshire. Earthworks consisted of broad, straight paths (or narrow rectangles) that appear to manifest alignments to the horizon and demarcate significant lengths delineated by ditches and sometimes punctuated by henges with concentric rings.

The new impulse toward building in stone arrived from the Preseli Hills of Wales, when a culture congruent with the megalithic at Carnac had arrived from Ireland. They used the local bluestone granite to build portal dolmens, with long counted lengths in between and at least one stone circle, which became Stonehenge 1. This led me to the view that the Stonehenge monument was the work of at least three megalithic regions, the native Cursus culture before 3000 BCE, the Carnac culture via Ireland before 3000 BCE, and the Mediterranean culture around 2600 BCE. The timeline for the different centers would then be:

1. **In England:** After 5000 BCE, the earliest megalithic tradition on the Atlantic coast was astronomical, located at 47.8 degrees north at Morbihan and concentrated near the village of Carnac, in Brittany. At Carnac, the solstice extremes of the sun from east and west on the horizon were at the angle of a {3 4 5} triangle, and the maximum and minimum standstill of the Moon aligned to the diagonals of the single and the double square.[1]

2. **In Brittany, Ireland & Wales:** At latitude 52 degrees, the Preseli culture was undoubtedly influenced by Brittany and Ireland, directly across the Irish Sea. Irish cairn art shares aspects of Carnac's geometrical art style seen most clearly in the Gavrinis cairn, a style resembling other goddess cultures but there clearly recording astronomical time phenomena. The Preseli Hills stand at exactly 52 degrees north, one degree of latitude north of Stonehenge, which itself stands exactly ¼ of a degree below the grand Avebury henge, this degree of latitude being the mean degree of the Earth,* as if it were a sphere that did not spin, associated with the spiritual nature of the Earth and created spiritual centers.

(*cont. from p. 45*) in Suffolk, the Cleaven Dyke in Perthshire, and the Dorset cursus." From "Cursus," Wikipedia, February 23, 2022. See also Roy Loveday, *Inscribed across the Landscape*.

*This means the north-south width of the parallel of latitude of 51 degrees to 52 degrees equals the degree distance of every degree *were the Earth spherical;* that is, of its mean size rather than oblated by its daily rotation causing days.

3. **In the Mediterranean:** Malta, at latitude 36 degrees north, had evolved the use of shaped sedimentary rocks, including linteled trilithons opening to rounded room plans similar to Sardinia and NE Spain (dated to 5000 BCE[2]). It probably created many of the Greek star constellations we use today, and it was succeeded by the Bronze Age Minoan civilization on Crete (see chapter 5). Crete had contacts with the ancient Near East and in south east Mediterranean, with Egypt, in the first half of the third millennium BCE, then the first dynastic period of pharaonic culture who surveyed the Earth's different latitudes, which allowed the Great Pyramid to be a sophisticated rationalization of the Earth that transformed megalithic metrology in the process. The model of the Earth written into the Great Pyramid's design was then embodied in Stonehenge 3, but in a circular fashion rather than as a pyramidal model of the Northern Hemisphere, its seven sacred latitudes and latitudes of Egypt.

Around 3100 BCE, the first phases of Stonehenge 1 and 2 introduced an evenly spaced stone-circle geometry, intentionally sited, as stated, within the Earth's mean parallel of a latitude between 51 to 52 degrees north, using bluestones from the Preseli Hills and design elements evidently developed by the Preseli culture in western Wales, exactly on the north of that latitude. The stones were quarried from the Preseli Hills, and perhaps assembled into a stone circle first before being transported to the largely empty site of Stonehenge, exactly one-quarter of a degree south of the giant henge of Avebury. The first stone circle at Stonehenge is now called the Aubrey Circle. It has 56 sockets where 56 Preseli bluestones stood.* These stones were later reused to form rings and an ellipse and then a horseshoe when the site was redeveloped after 2600 BCE when Stonehenge 3 was built within the Aubrey Circle, using a completely different style of stone circle of local sarsen sandstones, formed using dressed uprights and horizontal lintels, to construct a continuous ring of 29 stones plus a half stone, supporting an annular ring of lintels to form 30 trilithon portals, in the round. This structure is thought to have been the Hyperborean "Temple

---

*Robin Heath has shown that the Preseli bluestones led to the Aubrey Circle at Stonehenge. He has even shown that a metrological geometry of a Pythagorean triangle {5, 12, 13} links the source of the stones with the Stonehenge monument, as the two ends of the long side of 13 with the 12 side running from west to east between the longitude of Preseli and that of Stonehenge. See Robin Heath, *Proto Stonehenge in Wales*.

of Apollo," known to classical commentators. Within that circle, an ellipse of five taller trilithons were constructed. Stonehenge 3 was therefore incongruent in style to the earlier Stonehenge 1 circle, and, arguably, the sarsen circle and trilithon ellipse within it are therefore a somewhat mysterious double monument, synthesizing the megalithic traditions of that epoch (see figure 3.1).

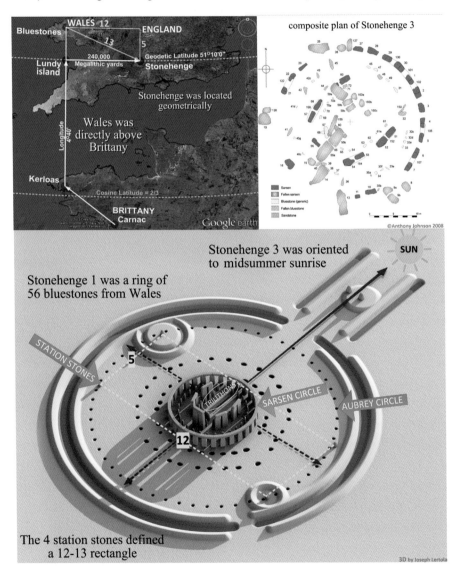

FIGURE 3.1. *Top left,* geographical relationship among Carnac, the bluestones, and Stonehenge; *top right,* a present-day Stonehenge 3 site plan; *bottom,* three-dimensional render of the Stonehenge 1 and 3 sarsen monuments with annotations. Note the identical 5-12 rectangle between the bluestones from Wales and Stonehenge in England and the four station stones.

The summer solstice sun shone into the new central horseshoe of five trilithons, which was itself within the new sarsen ring of 30 uprights. Meanwhile,
the 56 Preseli bluestones from the outer circle were reformed into a circle within
the new sarsen ring and a further horseshoe within the five trilithons, leaving
56 empty sockets in the chalk known today as the Aubrey Circle. The difference between the bluestones and the sarsens, quarried 40 miles north, could be
explained if Stonehenge 3 was built by people from the Mediterranean using
the local sandstone. For example, the lintels remind us of Malta and the stone
joints, of those used in Egyptian sacred building of the period including the
pyramids.

THE BLUESTONE CULTURE

Bluestones are a relatively rare variation of granite found in the Preseli Hills
of western Wales. The Preselis are formed from a remarkable set of volcanic
stubs whose main ridgeline moves exactly east–west and thus demarks a glacial
upland region between the hills and the northern coast, closed off to the west
by Carn Ingli (angel mountain). The winter sun therefore sets in the southwestern dogleg of the hills, and the whole area has a scattering of megaliths left over
from what is now a farming district focused on sheep and cattle, though wild
horses roam the hills that are now a national park.

The monuments include Pentre Ifan, the largest portal dolmen in Europe,
whose age is said to be 5000 BCE. Robin Heath developed an interpretation for
the locality as a single landscape temple of monuments related through alignments to the sky, expressing a kind of zodiac to the mountains centered on a
large, egg-shaped earthwork called Castell Mawr and, west of this, a very large
geometry of interlinked equilateral triangles,[3] out of which I selected three to
make figure 3.3. The equilateral triangles are the straight-line components of
the vesica geometry of overlapping circles; when vesicas are linearized into triangles, six of them can form a hexagon fitting within a circle. It is thus that two
such hexagons made of equilateral triangles can find eight of their equilateral
triangles shared in common.

At Göbekli Tepe the equilateral triangle linking enclosures B, C, and D,
expressing (as it does) the square root of 3 ($\sqrt{3} = 1.732$ in the ratio between
height and half-base), allowed the Moon's nodal period to be related to the
Jupiter orbit of 4332 days, when scaled down by the solar excess over the lunar
year to 3929. The same geometry was achieved in western Wales, but on a far

FIGURE 3.2. Pentre Ifan is Europe's largest dolmen. Note the volcanic "cairn" visible in the distance beneath its capstone.

larger scale, by using feet and yards, than at Göbekli Tepe, but still achieved by the ratio 400/231 equaling √3.

In figure 3.3, half of an equilateral triangle exists between the large portal dolmen of Pentre Ifan, a smaller portal dolmen in the north called Llech y Drybedd, and an old "fort" to the east, at the eastern base of Carn Ingli.

In figure 3.3, the reference length of 11,778 feet, multiplied by √3, equals 3 × 6800 feet, the number of days in the nodal cycle, implying the units used were yards of 3 feet. And dividing 11,778 feet by 3, one obtains 3926 yards, the number found at Göbekli Tepe when the half base of 231 is multiplied by 17 to equal 3927. This large structure on the landscape is therefore 20,400 feet (6800 yards on the ground). The geometry was a megalithic version of the same data found in the equilateral triangle at Göbekli Tepe, but there it was numerically miniaturized by division by 17 and the use of the unit of 12/77 feet. This implies that the smaller version was a preformed

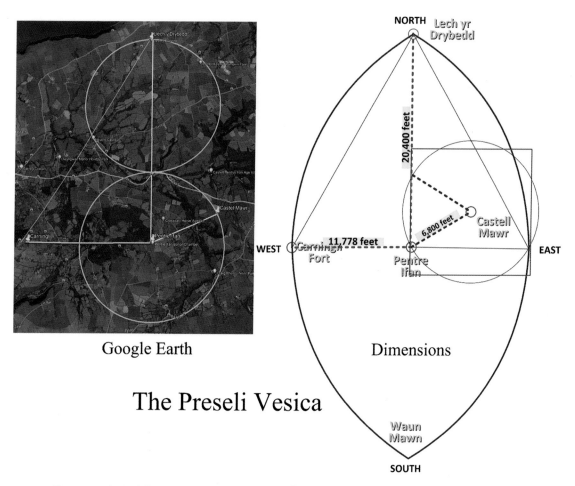

Google Earth   Dimensions

The Preseli Vesica

FIGURE 3.3. My own interpretation adapts Robin Heath's Preseli Vesica to show 3 × 6800 feet vertically from Pentre Ifan to Llech y Drybedd, which equals 20,400 feet and which, divided by √3, equals 11,778 feet to Carn Ingli Fort. It is also significant that an equal-area geometry, when centered on the Castell Mawr circle, is 6,800 feet from Pentre Ifan—the number of days in the nodal period of 18.618 years—and the square of that geometry is then the 33 years of the solar hero, with side equal to 12,053 day-feet or 33 years.

legominism* built into enclosures B, C, and D, while the Preseli version was the result of an actual astronomical count—requiring a far larger size for practical reasons.

Legominism was coined by G. I. Gurdjieff, who, while searching for the ancient world, found many hidden patterns in cultural artifacts where knowledge had been subtly hidden and preserved for future generations to recover, in buildings but also music, songs, images, and so on.

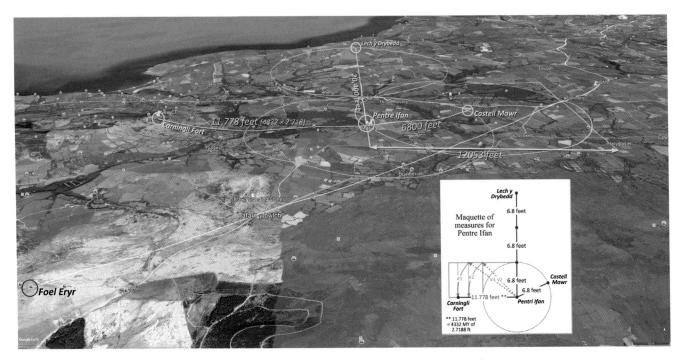

FIGURE 3.4. The Preseli uplands from the south, showing the geometry from figure 3.3. Carn Ingli is to the left, Pentre Ifan is central, and the raised observatory of Castell Mawr is center of a circular geometry of radius 6800 feet. Near the sea in the north is Lech y Drybedd (pron. *Dribeth*).

The primary drive for megalithic cultures, given their whole-number numeracy, was their search for a whole-number solution, and here the triangle was accurately rationalized as 20,400/11,778, yielding 1.73204, an accurate √3 to one part in 216,075! That 3 times the nodal period (20,400) should have a rational partner (11778) to "solve" √3 is unlikely enough, but 11,778 can decompose in very interesting ways such as its factors (2 × 3 × 13 × 151), 4332 × 2.71875, or 361 × 10.876, where 4332 days is the orbit of Jupiter in days and its factors 12 × 361 include the number of days (361) for Jupiter to transit a single sign of the zodiac. In addition, 2.7188 feet is the proto-astronomical yard, seen generated at the Le Manio Quadrilateral, while 10.875 is one-third of that yard, in day-inches.[4]

The geometry seems to re-present a discovery: that the Moon's nodal period of 6800 days is somehow the synchronous result of Jupiter's orbital period and the excess of the solar year over the lunar year. That two cosmic periods, multiplied by √3, should yield a rational and accurate product, 6800 (the nodal period in days), is an astounding result, here multiplied by 3 to equal the 20,400 feet to Lech y Drybedd. That √3 relates the nodal period to Jupiter and the Sun, in any way, is very unlikely unless there actually is such a relationship that can be portrayed geometrically as an effect of Jupiter and the Sun upon the lunar orbit, *naturally* organized within the equal area model. And because this same usage for the equilateral triangle has already been interpreted at Göbekli Tepe (in chapter 2) but to a different scale suggests this was part of an astronomical corpus lost in deep time. Its physical causation may be due to a form of Kepler's area law for planetary orbits when applied to our planet's moon.

This nodal relationship to Jupiter and the Sun must relate to Jupiter's resonant lock with the lunar month (29.53 days), whose phases of illumination are the combined result of the differential motion of the Sun and Moon during the Moon's orbit (27.32 days), so that the synod of Jupiter (398.88 days) is 13.5 lunar months, and 13.5/12 lunar months is, remarkably, the musical whole tone of 9/8.

But how could this nodal relationship have been discovered? Probably working from the time between lunar maximum standstills of (rationally) 6800 days (and *not* as a fraction of 18.618 years, at that stage). If, by fluke, the orbit of Jupiter was then being quantified by the megalithic astronomers using megalithic yards of 2.71875 feet, then 4332 days would become 11,777.625 feet, a length that could be seen (in feet) to be 1.732 of 6800 day-feet; that is, by first having two measurements, the nodal period of 6800 day-feet, and then the Jupiter orbital period of 4332 day-feet, while using megalithic yards the lengths compared geometrically, in feet, would give the geometrical solution in yards *exactly,* because yards relate to megalithic yards as 36/32.625 = 32/29. This would then be seen as the joint influence of Jupiter and solar excess (the Sun or solar hero) over the nodal period—by an absurd happenstance or extreme good fortune.

But if the concise legominism of the equilateral was already demonstrated at Göbekli Tepe, why is it being reenacted as a discovery in the Preseli Hills? Perhaps it was a genuine rediscovery of knowledge belonging to the Mesolithic tradition of astronomy, survivors of an Arctic center (chapter 8) destroyed by the freezing of the Arctic Sea, and appearing in miniature at Göbekli Tepe but full-size in the Preseli Hills. In that case, the variation of the invariance of the sky at lower latitudes could have been a new megalithic study of astronomy at different latitudes, from which a survey of the Earth's shape, as non-spherical, was quantified.

THE SIGNIFICANCE OF LATITUDE
FOR THE MEGALITHIC

In the Northern Hemisphere, the Sun rises higher as one travels south upon the surface of the Earth, but a vertical pole's shadow could not accurately establish latitude. Fortunately, stars touch the southern meridian at a fixed altitude so that bright stars could be catalogued according to their angle at one site so that, in moving toward the North Pole, one would seek an exact degree in a bright star's "declination" to know you had traveled a degree in latitude; that is, it is possible to measure the distance of a single degree of latitude between sites north or south, using the change in declination of a fixed star. Such parallels of latitude grow longer to the north and shorter to the south; that is, the Earth seemed to have a non-spherical shape, and, from this, the polar radius of the Earth was less than it's equatorial radius. And it was at the latitude of Preseli and Stonehenge—which still retained the average, mean length for a degree—that a spherical Earth would have. To understand this means that the perfect geometry of circle and sphere needed to morph the Earth into a shaped meridian and circular bands of latitude whose angle to the stars was their latitude.

What Sir Isaac Newton called the "oblate spheroid"—an elliptically shaped Earth—has the polar radius shrinking relative to an enlarged equatorial bulge. The polar radius of the Earth and its intermediate mean radius are in the rational ratio of 441 (the mean radius) to 440 (the polar radius). Newton believed that the ancient metrology of the Jews, biblically gleaned by Moses from Egypt, implicitly held the magnitude of the polar radius in terms of, for example, the Royal foot of 8/7 feet and sacred double-foot rod of the Jews of 72/35 feet.* If he could find the polar radius he could calculate the distance, mass, and size of the Moon, according to his new law of gravity. Returning to the ancient units of earlier megalithic science, 441/440 was identified as the ratio, determined because the equatorial radius could be estimated from a given latitude and a known portion of that latitude's angular extent.†

If the megalithic had evolved astronomical land-surveying methods, a new knowledge of the Earth's size and shape would emerge. The equal-perimeter model of the Earth and Moon was an approximation whose discovery must be

---

*Sometimes called a polar foot, because it divides into the polar radius, but now resolved as the root common Greek foot of 36/35 feet, two of which make the Sacred Rod of the Hebrews as 72/35 = 2.0$\overline{571428}$ feet.

†As discussed in my *Sacred Number and the Lords of Time,* fig. 1.1 and chapter 5.

lost in the Ice Age or before. Based on a circular meridian, monument build-ing retained its circular form and dimensional numbers as if the mean Earth was a kind of blessed land, upon which was superimposed the existential Earth that was spinning and hence oblate. However, it would appear the Egyptians found a simple way to describe the Earth using the three key radii—the polar, mean, and equatorial—as accurately as any modern generalization. This uni-fied geodetic model of the Earth became the theme for both the Great Pyramid and Stonehenge as repositories of cosmological standard lengths, within the same century. Though each way of embedding the model was different (pyra-midal and circular), the model behind both buildings was identical, implying that whoever designed them had access to the same model. This new knowl-edge can explain the overarching meaning of the Great Pyramid of Giza—a building contemporaneous with the Sarsen Circle—as being a representation of the Northern Hemisphere.[5] Its actual height is 440 Sumerian feet, while its full height, were it to come to a point, would be 441 Sumerian feet. Its height is therefore the polar radius relative to the mean radius the Earth *would have* were it a perfect gravitational sphere* that did not spin. This was shown differ-ently in the design of the Sarsen Circle, but here we shall discuss the basics of how the Great Pyramid was built to represent the Earth and its different lati-tudes, according to key parallels of latitude established by a global survey from Ethiopia to Europe, as defined by megalithic observation using 360 degrees in a circle, the same method as the Sumerians used and that we still use today.

A PYRAMIDION FOR THE GREAT PYRAMID

The aim here is to see the results of what we assume to have been a geodetic survey of different latitudes directed by the earliest dynastic Egyptians, with support from the Atlantic megalithic regions, regions closer to the pole, includ-ing the mean degree. South of Egypt lay other key degrees that any survey must take into account to come up with accurate results, for example, in Ethiopia. We can infer that the Egyptians evolved the worldwide system of land measure-ment *because* the Great Pyramid of Giza encoded different key latitudes of the Nile Valley as well as of the globe, using variations in height multiplied by its side lengths.

The difference between the 440-foot height of the pyramid and its

---

*Planets over a certain mass become a near perfect sphere due to their own gravity.

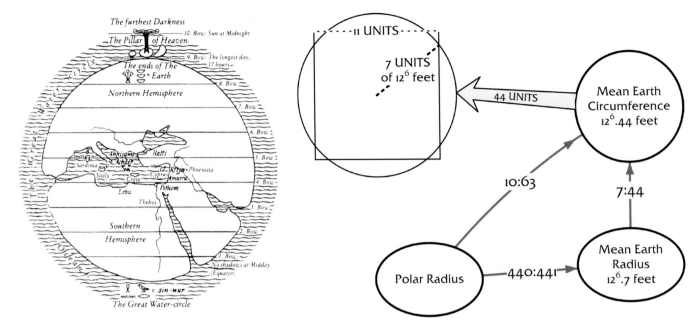

FIGURE 3.5. *Left,* a map showing the Egyptian view of the world by 1200 BCE, before the end of the Bronze Age, showing nine bows (our "parallels") of key latitudes, the bow numbers 4 to 9—including the Nile Delta, Delphi, southern Britain, and Iceland—a map based on an ancient geodetic profundity (*right*) that the Earth's dimensions could be rationalized according to three approximations of pi: accurate as 22/7, overlarge as 63/10 or 6.3 and too small as 25/8, used in the microvariations of 175th of a foot.

theoretical 441-foot height is 1 Sumerian foot. Small models of pyramids, called pyramidions, were often used as scale models of a whole pyramid. The Giza pyramidion appears lost forever, but some others have been found and identified for a number of other pyramids. The exact form of the Great Pyramid's pyramidion is therefore a subject of conjecture we seek to resolve here.*

The Pyramid's Use of Rectangular Numbers

As just stated, John Neal demonstrated that the 480-foot height of the Great Pyramid, when multiplied by its four *unequal* sides, formed four rectangular *areas* (in square feet) whose number is the *linear* extent (in feet) of four latitu-

---

*This whole matter was only made possible owing to the work of John Neal, whose *All Done with Mirrors* showed that the pyramid's four sides were in proportion to four latitudes through which the Nile passes. The sides in feet, multiplied by its height of 480 feet, gave areas equal to the exact lengths of these latitudes, but *only if feet were employed.*

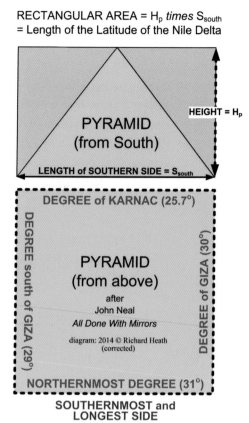

**RECTANGULAR AREA = H$_p$ *times* S$_{south}$
= Length of the Latitude of the Nile Delta**

**PYRAMID
(from South)**

HEIGHT = H$_p$

LENGTH of SOUTHERN SIDE = S$_{south}$

DEGREE of KARNAC (25.7°)

DEGREE south of GIZA (29°)

DEGREE of GIZA (30°)

**PYRAMID
(from above)**

after
John Neal
All Done With Mirrors

diagram: 2014 © Richard Heath
(corrected)

NORTHERNMOST DEGREE (31°)

**SOUTHERNMOST and
LONGEST SIDE**

FIGURE 3.6. The Great Pyramid's modeling of key latitudes of the Nile using rectangular numbers between its height and four different side lengths. After John Neal, *All Done with Mirrors.*

dinal degree lengths, in feet along the Nile valley (of its Delta, of Giza, south of Giza and of Karnak).

In the case of the pyramidion for the Great Pyramid, the southern "index" side is rational in feet as 756 feet. This length, multiplied by the pyramid height of 480 feet, would form the latitudinal degree distance* for Ethiopia, a prime location for any geodetic survey. I then saw that the pyramidion could have sat on the truncated top of the pyramid and taken a form suitable to give rectangular areas equal to the four key latitudes of the Nile delta, Athens, and Delphi in Greece, Avebury and Stonehenge (at the mean latitude) and even of northern Iceland.

---

*The degree distance is taken here as meaning "the width of a parallel of latitude between two lines of latitude one degree apart." The mean degree distance is the width of the 52nd parallel, defined as between 51 and 52 degrees of latitude. Parallels of latitude confusingly run east–west, each successive parallel gets wider, traveling north. One must remember that the first parallel is from the equator to the latitude of 1 degree. Some argue that the parallel is half a degree to either side, but that is not the case.

A pyramidion for the Great Pyramid, expressing the key latitudes of the Earth, as stepped pyramids & ziggurats did.

extending the geodetic ideas of John Neal
© 2020 Richard Heath: sacred.numbersciences.org

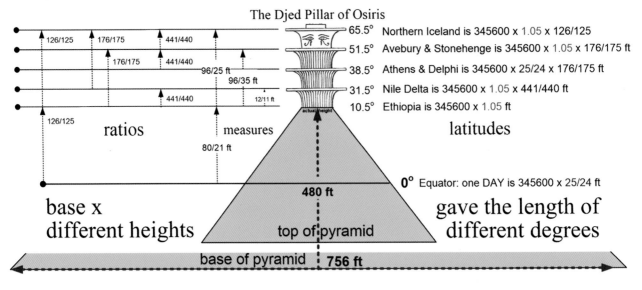

The Djed Pillar of Osiris

- 65.5° Northern Iceland is 345600 x 1.05 x 126/125
- 51.5° Avebury & Stonehenge is 345600 x 1.05 x 176/175 ft
- 38.5° Athens & Delphi is 345600 x 25/24 x 176/175 ft
- 31.5° Nile Delta is 345600 x 1.05 x 441/440 ft
- 10.5° Ethiopia is 345600 x 1.05 ft

126/125   176/175   441/440

176/175   441/440

96/25 ft
96/35 ft

441/440   12/11 ft

126/125

ratios   measures   latitudes

80/21 ft

actual height

480 ft

756 ft

0° Equator: one DAY is 345600 x 25/24 ft

base x different heights   top of pyramid   **gave the length of different degrees**

base of pyramid

FIGURE 3.7. Using a pyramidion to define the lengths of key latitudes on the globe, shown in fig. 3.5 and further quantified in fig. 3.8.

To illustrate how this might have worked, the pyramidion shown in figure 3.7 uses the djed column of Osiris to illustrate in principle that the existing pyramid could be used this way to access the degree distances of key latitudes.

The above overview shows the pyramid's function as a repository of geodetic knowledge gleaned from a survey where seven key latitudes were defined, with the mean Earth radius becoming a symbol of whatever had transformed the Earth into this rational profile, or geoid. The mean radius as 882/800 of the mean ratio, then gives a good approximation of the equator, as 883 of the same units.

Figure 3.8 shows the rational differences between all the seven key degree distances In three cases, this equals 441/440, the ratio of the mean radius to the polar radius. Other ratios of 176/175 and 126/125 made other latitudinal ratios rational, 126/125 then being 176/175 × 441/440.[6] Thus, the *differences between the key degree distances* (as well as the ratio between three different Earth radii as 441/440) gave the link between latitudes, ancient metrology, and the Great Pyramid—440 units high to its ideal apex of 441—and the Earth itself in key-radii and seven-latitude framework. The pyramid was a computational synthesis.

These ratios were therefore expressed by the metrology built into the

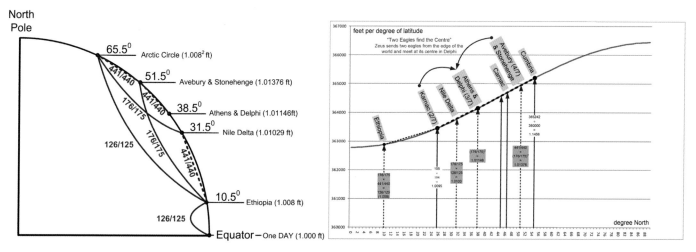

FIGURE 3.8. *Left,* a cross section of a single meridian and key latitudes emerging from rational ratios. (After John Neal, 2000) *Right,* graph of increasing degree distances from equator to North Pole, and, below that, key latitudes of megalithic and ancient-world significance.

pyramid but then also in the area products between its pyramidion's different heights, that became a model of the latitudinal shape of the Earth, its geoid. The pyramid also expressed π as approximated in the equal-perimeter model but again, also between those key latitudes of figure 3.5 (*left*), where each ratio shown below has the product of two rational approximations to π:

441/440 is 63/10 × 7/44,
176/175 is 4/25 × 44/7, and
126/125 is therefore 63/10 × 4/25.

In this way, the differences between the latitudes became integral to the metrological system of different foot ratios, becoming the micro-variations found within historical metrology in its different modules. While these micro-variations, including 176/175, were clearly used by the megalithic culture in southern Britain, they were used to maintain whole numbers, of slightly varied feet, between a circle's radius and its perimeter, in length. And this is the facility more simply provided by the 12/77-foot subunit found in the Mediterranean tradition. From this one can deduce that a search had been made, by the epoch of the Great Pyramid's construction, for those specific ratios that conformed to the value of π beyond the simplest of 22/7. That search revealed more approximations to pi to create the global concept of ancient metrology so that the Earth and its monuments should both be intelligible, by whatever means and at any scale, using only rational fractions.

BRITAIN'S CIRCULAR GEODETIC MONUMENT

The Sarsen Circle lies ¼ of a degree south of Avebury Henge, whose latitude is 4/7 of 90 = 360/7 (51.$\overline{428571}$) degrees north, and the linear distance between these latitudes is 86,400 feet, using the Manx geographical foot of 1.056 feet. The Sarsen Circle and the Avebury Henge therefore became a compound monument related to the size of the mean earth through the distance between their latitudes. Yet Avebury was a henge from the preceding Cursus culture and the Sarsen Circle was built within Stonehenge 1. The latter was already ¼ degree south in latitude and so Stonehenge 3 marked a previous compound monument organized during a geodetic survey of mainland Britain, when Stonehenge 1 was constructed by 3000 BCE. This implies that when Stonehenge 3 was built around 2600 BCE, the same date as the Great Pyramid of Giza, determined by historical and carbon dating methods.

If Britain was a partner in the Egyptian geodetic survey, then the compound monument was part of the geodetic survey that expanded the megalithic metrology into a purely numerical system, where the more detailed figures for the key radii of the Earth were embodied in the metrological system itself, rather than, and less accurately, in the equal perimeter model. For example, the mean radius according to ancient metrology was 7 times 12^6 feet, which is 3958.$\overline{690}$ miles whereas that radius in the equal perimeter model was 3960 miles. The difference between them is is 3024/3025 and, if 3960 is multiplied by Fibonacci's 864/275,* the 3960 mile figure gives the same figure as 3958.$\overline{690}$ mile radius, for the circumference of the mean Earth, of 44×12^6 feet, or 24883.2 miles. Fibonacci's 864/275 is a more accurate π so that the equal perimeter model still applies in a more important sense, in that it is probably the surface area of the Earth that enables it, with the Moon and other factors, to be a living planet.

Stonehenge 3 is visually an orphan of style, though Robin Heath has noted the Sarsen Circle is in the ratio 7 to 19 relative to the Aubrey Circle's bluestones. Its masonry is not bluestone, but rather sandstone, which could be sculpted. Also,

---

*This fact was first realized by John Michell in his *How the World is Made*. He was perhaps trying to adapt the equal perimeter model to the metrological one.

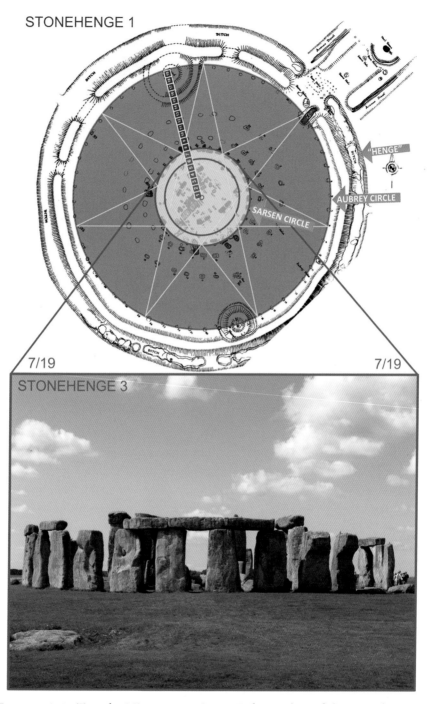

FIGURE 3.9. *Top,* the Nineteen to Seven Relationship of the Stonehenge 1 (bluestone) Monument and Stonehenge 3 (sarsen) monument, where 7/19 (0.368) of a lunar month is the solar year's excess over the lunar year. *Bottom,* the Sarsen Circle (photo: Gareth Wiscombe for Wikipedia).

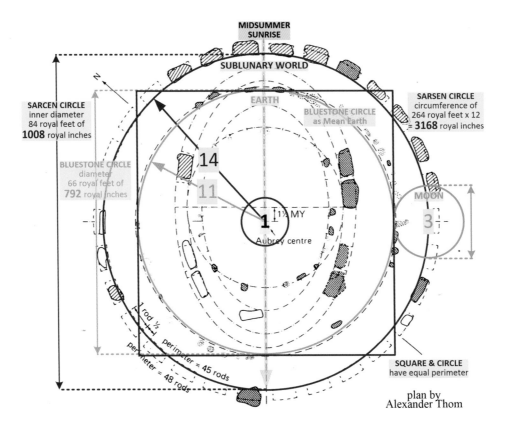

FIGURE 3.10. Stonehenge's raised lintel ring has 29.5 sandstone sarsens for the days of the lunar month. The ring's circumference and the perimeter of the square around the bluestone circle are both 3168 Royal inches. The inner circle's diameter represented the 7920 mile diameter of the Earth. The diameter of the sarsen circle was 1008, larger by 216 Royal inches, and represented the 2160 mile lunar diameter. In practice, Royal inches of 8/7 inches were enlarged by 3168/3125 = 1.01376, a "Pythagorean comma" that scaled the monument to match its 51.5° latitudinal length shown in figure 3.8 on p. 59.

the henge-like earthwork of ditch and bank surrounding the Stonehenge site is the reverse of a true henge like Avebury, and the Station Stone rectangle, within the Aubrey bluestones, appears to avoid the Sarsen Circle of Stonehenge 3, by design. I therefore propose that Stonehenge marks the site where a unique meeting took place of megalithic cultures from different latitudes who had been creating a model of the Earth. What started at Stonehenge 1 traveled north to measure Britain, while others went to the Mediterranean and Egypt, a work of many decades or even a century or two, culminating in the building of Stonehenge 3.

A further use for any Stone Age monument, after its original purpose, was as a repository for new knowledge and understanding, since the oral tradition had no other means to transmit such knowledge except through works that contained their high knowledge of the sky and the Earth, using geometry and numbers as lengths.

Many of the key dimensions of Stonehenge are well defined and, from this, the Sarsen Circle, can be seen as a geodetic and metrological temple, a "repository of standards"* with differential allusions to horizon alignments and day-counting astronomy.[8] The focal feature is a continuous ring of lintels that are exactly encoded, using metrological scaling, to the principal radii of the Earth. This sort of precision in stonework is not expected in Britain but very characteristic of Egypt, whose dynastic state structures were underwritten by the fecundity of the Nile. The elliptical ring of five trilithons is more reminiscent of Mediterranean astronomy and the Maltese temples, pointing to the synodic periods of Venus (see next section).

The stone circles of megalithic Britain were a clear break with the earlier use of earthworks, henges, and barrows, and, while made of large stones, Stonehenge 3 was unlike any other British stone circle, but in some ways similar to Malta. In an age of rude-stone—but nevertheless precise—circles, many not exactly circular and Preseli-like designs, including egg-shaped and flattened circles, followed. It is as if the work in Brittany, Ireland, and Preseli continued on in its own style, but enriched, after Stonehenge 3 was built, through blending with the traditions found in the Mediterranean. If the goddess culture was the original social background that could support megalithism, then the British megalithic would have followed the Paleolithic goddess into the Bronze Age. Stonehenge 3 could therefore mark a reunion with European megalithism after the geodetic work had successfully surveyed the size and shape of the Earth. It is for this reason that the Sarsen Circle came to express Egyptian geodetics (see fig. 3.10) and the mysterious five trilithons within it expressed Venus, the goddess of the inner solar system, in an ellipse of exact proportions, complementing the Sarsen Circle as both model of the outer orbit of the Earth, in the center of the local site at Stonehenge, where the solstice Sun shines at midsummer. Proceeding in this way, trilithons represent the goddess culture as being a model of the synod of Venus, who is the younger face of the triple goddess, latinized as

---

*This is John Michell's form of words, which motivated his search for the metrology within Stonehenge and the ancient world, reported in his "Ancient Metrology" booklet, published in 1981.

a girly Aphrodite but actually the sterner matriarchal goddess Ishtar, queen bee of the ancient Near East.

THE STONEHENGE TRILITHONS AS SYNODS OF VENUS

Inside the sarsen ring, there once stood a group of five trilithons, each made up of two uprights and a lintel stone, hence repeating the unique style of building found in the Sarsen ring itself. Higher than the Sarsens, the trilithons partially punctuate an ellipse open toward the midsummer sunrise, the axis of Stonehenge and its solstice-marking "heel" stone. (Compare fig. 3.11 with 3.12, where the former is looking north in the latter.)

The Horns of Venus

The symbolism of Venus therefore involved the sun, and the number 5 (the original number of trilithons), while these expressed something involving five close pairs. The dominant astronomical significance of the number 5 comes through in the brightest planetary phenomenon of all, when the planet Venus approaches the Earth. As Venus approaches from the east, it precedes sunrise in the evening sky. It is often therefore called the evening star. Venus then shoots

FIGURE 3.11. On the *left* are two of the trilithons (within the Sarsen Circle to the *right*). Both monuments use lintels, the inner monument joining just a pair of stones and second unify all 30 stones into a raised continuous ring of lintels.
Photo: Bernard Gagnon for Wikipedia Commons.

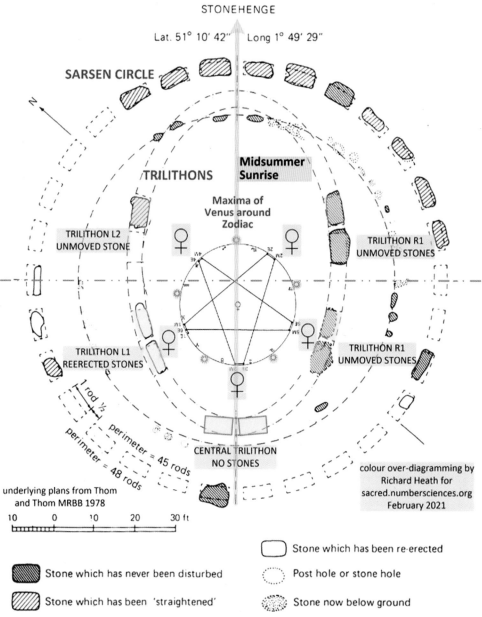

FIG. 11.5. The sarsen ring and its interior, with geometry imposed.

FIGURE 3.12. The five trilithons of Stonehenge 3, highlighted in yellow within the Sarsen ring to express the five evening and morning star couplets that occur in eight practical years of 365 days. Plan from Thom and Thom, *Megalithic Remains in Britain and Brittany* Central portion is figure 3 upside down to match the horseshoe of trilithons.

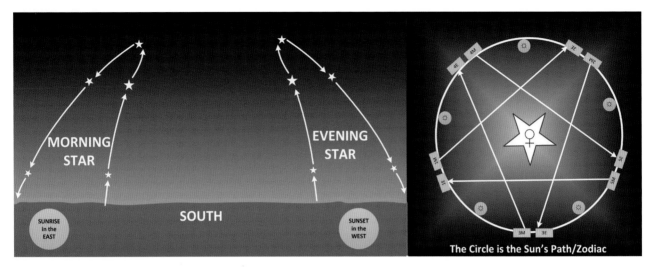

FIGURE 3.13. *Left,* the Horns of Venus when the evening and morning "stars" are visualized over an extended present moment during the Venus synodic period which is the recurring time cycle of her phenomena, seen from the earth, lasting 1.6 (or ⅝) practical years of 365 days. Adapted from my *Matrix of Creation,* fig. 2.2. *Right,* the horns of evening (labeled *E*) and morning (labeled *M*) stars shown upon the background circle of the zodiac, each successive pair being 3/5 advanced within the solar year. (after Joachim Schultz, *Movement and Rhythms of the Stars,* fig. 88.)

past the Sun and reappears in the morning sky, again growing in brightness as the morning star (see fig. 3.13).

Through counting, the megalithic observers had found that a practical year of 365 whole days (5 × 73), formed a count 5/8 of the synodic cycle of Venus, which was 8 × 73, or 584 days. It was also obvious that the horns of the evening and morning star in the sky were bracketing the Sun, just as the elliptical cup of the five trilithons erected at Stonehenge, bracketed the solstitial Sun; a sun that travels every day from east to west while advancing east by one day *in angle* (by definition) each day.

If the Earth was their viewpoint, then the Sun's path over the year could, like the Sarsen Circle, be seen as a circle of 365 days and, when the time between evening or morning stars was counted, the result was 584 days between the successive hornlike (and brilliant) manifestations of Venus, as with the loops of the outer planets. The count of 584 days is 219 more than 365 days. The Sun has therefore moved 3/5 of a year forward and hence it became noticeable, as stated above (that 1/5 of the practical year is 73 days, the practical year 5 units of

73 days long while the Venus synod is 8 units of 73 days long). The Venus synod therefore has exactly 1.6 (the Fibonacci ratio of 8/5) practical years between its phenomena.

The movement of Venus across the zodiac therefore describes a five-pointed star, emblematic of the goddess. The number 5 and all of its properties are synonymous, by association, with the planet Venus, who, as noted above, became the leading and youngest of the triple goddesses of the ancient Near East, and a role young women took when queen. The golden proportion or mean (1.618034 . . .), often seen in classical and neoclassical architecture, has the number 5 as its geometrical root. Also, many living bodies share forms derived from the number 5, or from numbers in the Fibonacci sequence that approximate the golden mean.

The Fibonacci series {1, 1, 2, 3, 5, 8, 13, 21, . . . } has successive numbers that sum to give the next number, and each new ratio, between successive numbers in the series, yields an ever-better approximation to the golden mean: {2, 1.5, 1.666, 1.6, 1.625, 1.615, . . .}.

Almost certainly, therefore, the general form of the horseshoe of trilithons at Stonehenge appears to have monumentalized Venus, Fibonacci numbers, the golden mean, and its related 5-based geometries and the Sun. This can now be confirmed by looking at the metrology used in building the "horseshoe" of trilithons, which was simple yet indicative of 8/5, when seen in units of the megalithic yard* used at Stonehenge. Stonehenge phase 3 is widely found to have used a megalithic yard (MY) between 2.72 to 2.722 feet long in its construction. Thom says:

> The intended width of the [remaining] uprights in the trilithon ring seems
> to have been 1 rod [equal to 2.5 MY], and the spaces ¼ rod [⅝ MY] and
> 4MY [10 units of ⅝ MY], all measured on the inside ellipse.[9]

That is, the common unit of measure of the inner faces of the trilithon uprights employed a Fibonacci ratio of the megalithic yard of 5/8 MY, which is the ratio of the practical year of 365 days relative to the Venus synod of 584 days. Therefore, each trilithon is a single Venus synod in a set of five: the width of each pair being 2 rods or 8 × 73 days and, after 5 synods of Venus,

---

*Any unit of length 2.7 to 2.73 feet long, after Alexander Thom discovered 2.72 feet and 2.722 feet as units in the geometry of the megalithic monuments of Britain and Brittany.

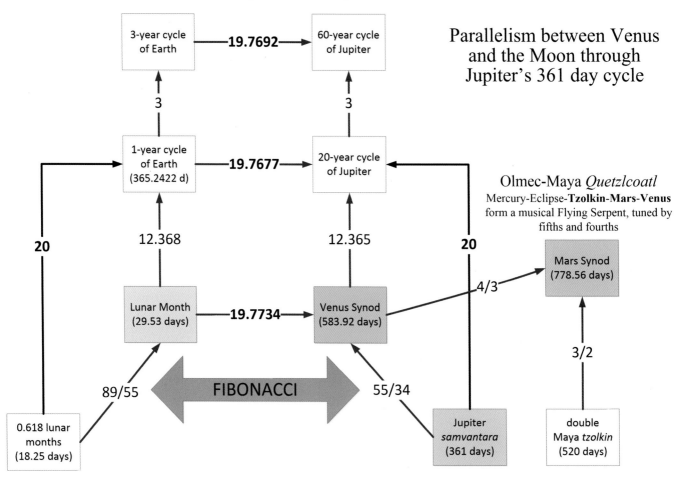

FIGURE 3.14. The five-synod process by which the numbers 5 and 8 return the horns of Venus to the same point on the zodiac after 8 practical years.

8 practical years have elapsed (40 periods of 73 days) and the cycle of Venus manifestations will then return to almost exactly the same place in the zodiac (see figure 3.14).

In other words, the front widths of all 10 trilithons, equal 40 units of 5/8 MY, which, seen as 73 days each, equals 8 practical years of 365 days, which in turn equals 5 Venus synods of 584 days. For more metrological detail, please see the box "The Metrology of the Trilithon Ellipse on page 69."

Metrology of the Trilithon Ellipse

Once it is seen that the central portion of Stonehenge related Venus to the Sun and Earth, a very practical symbolism emerges in which the trilithons are the double manifestation of Venus, as evening and morning star. There is a direct resonance between the practical year of 365 Earth days and the Venus synod, which is 365 × 8/5 = 584 days. The inner ellipse of the five trilithons has a bounding ratio of 5 to 8, the major axis

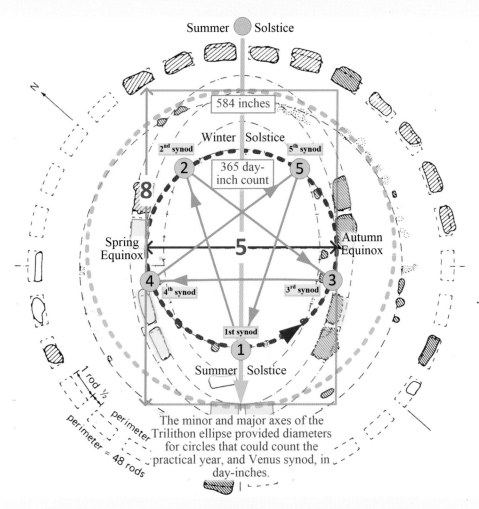

FIGURE 3.15. The minor and major axes of the Trilithon ellipse (ratio of 5 to 8) provided diameters for circles that could count the practical year (as 365 day-inches), and the Venus synod (as 584 day-inches). The eventual form of the trilithons naturally emerged from the counting process: a 5-fold star or pentacle of evening star and morning star periods.

towards the solstitial sun. Figure 3.15 shows that the minor elliptical axis of 5 units had a measured length that, used as a diameter, creates a 365 inch perimeter.

This suggests that counting could move one inch per day in a counterclockwise direction starting from the landing of the summer sun on the central trilithon's gap, perhaps when the evening star has appeared in midsummer. The practical year can be seen as made up of five periods of 73 days each, as after a year of five such periods three more remain before the advent of another evening star. As shown, the second synod lands 3/5ths into the next year, and this will repeat four more times to describe a pentacle within the circle, which represents the Zodiacal path of the sun in the year. The symbol of five trilithons would then be seen as the record of five evening and morning manifestations of Venus, a form then "opened" to face the midsummer sun.

The intuition that life comes from the Sun, into the womb-like and sacred space of the Earth, articulates the triple goddess. The Moon as "old lady" has a protective boundary of 29.5 stones whose inner circumference is 3168 royal inches. The Earth as "woman" is the bluestone ring of diameter 792 royal inches, each representing 10 miles. "Young maiden" Venus is symbolised as the pentacle of 5 synods in 8 years. The triple goddess receives from a male Sun the cosmic life principle, symbolized by the midsummer solstice sun.

The Modern Perspective

In modern gravitational thinking Venus and the earth can be seen as locked through their regular synodic proximity, exploiting the fact that a planet whose orbital period is $8/13$ that of the Earth will automatically have a synodic period of $8/5$ since 8 plus 5 equals 13. The $8/5$-year synod divided by $8/13$-year orbit of Venus = $13/5$ (2.6) orbits of Venus for $8/5$ (1.6 year) synod of Venus. Similar episodic conjunctions appear to have led to the many other number coincidences found within the geocentric astronomy of the megalithic period.

When the synod occurs, Venus returns to being the evening and morning star. For this to continue, the numbers {5, 8, 13} must be exact but also are kept exact by those integer numbers. Forces out of balance act according to the deviation from the mean, as in the cybernetic regulator in achieving homeostasis. Resonance represents a stable energy condition where any change in state will be pulled back into stability by one influence that would accelerate and another, decelerate a given orbital body.

PART TWO

THE MEDITERRANEAN TRANSMISSION

The history of the megalithic, including the anomaly that is Göbekli Tepe, was ended by a Bronze Age dominated by ancient Near East cultures ruled over by god-kings who benefited from agricultural surpluses and the trading networks for goods of all kinds. But in the Bronze Age the first civilization in Europe (now called Minoan) was on the island of Crete. The proximity of matriarchal Crete to matriarchal societies on the mainland and to the ingress of patriarchal influences from the Near East and the north gave Crete the opportunity to live on within the cultural memory of the later classical cultures of Greece, then Rome, and, eventually, early Islam and medieval Europe. Matriarchal traditions were sometimes admired and other times distorted to suit "the male gaze" or fears of powerful women. Without this transmission through myth and social habits, it would be even harder to understand this pivotal period between matriarchal societies and patriarchal ones, even when the latter behaved in a "matrilineal" way such as when a town's fields were allotted by gambling, given to the god, or used for competitive sport.

4

TIME, GENDER, AND
HUMAN HISTORY

The history of the ancient Near East, notable for its god-kings, cities, civilizations, empires, wars, and pyramids/ziggurats, seems dominated by men and hence largely patriarchal. At the same time, the Mediterranean had matrilineal tribes led by women, the social organization of which extended from the gathering of food, the raising of children, the domestication of animals, the establishment of households, and new laws of land ownership and allocation. The Mediterranean and northern Africa both have a deep history of matrilineal and matriarchal tribes.

The Maltese astronomers, for example, with their veneration of goddess figurines, and Bronze Age Crete's matriarchal art during much of the Minoan period are never to be joined-up facts for most patriarchal historians. The idea of women organizing societies became a patriarchal taboo largely through invasion of patriarchal cultural norms, suppressing mention of it or spreading horror stories about the domination by women as monsters such as the Gorgon, which had snakes for hair. If self-sufficient farming was practiced on the island of Malta, this gets merged into the "one size fits all" notion that if anyone farms, they are Neolithic, ignoring the skill of nomadic foragers or Stone Age gatherers cum horticulturists. When Evans made his great discovery at Knossos of the Minoan capital, having saved the site by buying the land to excavate there, he could not help but invoke the male King Midas of Greek myth to name the civilization, rather than recognize in the artifacts that it was the goddess who ruled as queen.

Applying the same rule of thumb, the Greek myths can be seen to have transposed women into Amazons or monsters, inadvertently recording their emancipation as leaders similar to some women of the twentieth century. The

73

myths were written or edited by the patriarchs who took over the matriarchal and matrilineal societies; that is, Greek myth remembered the women but were speaking through the men of their time, using largely forgotten symbols of matriarchy to provide warnings, lest the women make a comeback and again suppress kingship. The Gorgon was in fact the aegis, or crest, on the leather armor of some matriarchs, or a symbol of power within an accompanying bag. But in writing down the myths, the editors never imagined they were documenting the past for distant and unknown future readers. The literary criticism of myths has shown us that women once ruled many tribes, nations, and even Europe's first civilization, on the island of Crete. A blind spot therefore exists within our histories concerning gender leadership at this crucial juncture in history, preceding the classical civilizations.

GENDER SELECTION WITHIN PREHISTORY

In the Late Stone Age, group intelligence was *increased* by not being tied to the land. Complementary to this female organization, the male skill was of an environmental intelligence regarding the periphery of the group, for men had to move beyond the tribal boundaries to hunt animals in the hinterland, including the ranges of other tribes. Males of later times would obey tribal taboos, using animal and plant symbols (like the zodiac) to define whom they should not breed with. This meant that leadership, meaning-making, and the continuity of the tribe were provided by women, and individuals inherited their mother's, not their father's, name, in addition to a personal and collective tribal identifier.

The intelligence found in women and men has not changed in modernity, as both types of intelligence are enshrined in the sidedness of the human brain, the emotions, and physical bodies. But natural selection is a slow and blunt instrument in the social evolution of men and women, still adapted to the Stone Age and not a late Holocene. Communities and tribes have become ever more dispersed and then re-concentrated into patriarchal towns, cities, and nations because atomism works well when applied to economic exploitation and growth. Land ownership has been displaced from the tribe and put into the hands of private individuals, organizations, and the state, creating a land-owning class. These were perhaps the first people to ever vote, in the nascent democracy of Athens. And yet, their patroness Athena was a transformed goddess of matriarchal Greece or perhaps north Africa, who had emerged from Zeus's head "fully armed and with a shout," because she was a fighting goddess from when women fought to be the sacred matriarch or queen bee. She had been recycled to become goddess and anima of patriarchal learning and literacy (see fig. 6.6 on p. 125 for her parthenogenesis). The idea of fighting women can be seen in the story of the Amazons in Homer's *Odyssey,* set in the Black Sea in which an island of women bewitch the sailors into sexual captivity—one take on matriarchy. As leaders, it may be that some selective breeding occurred in matrilineal societies, as portrayed in Frank Herbert's *Dune* and the Bene Gesserit order of a female sisterhood.

Athena had accidentally killed her best friend, Pallas, in friendly combat, so taking her name as Pallas Athena. It is therefore no surprise that a matriarchal prehistory's achievements and mores became veiled within the Greek myths and forgotten as if unimportant. Gender leadership tipped indefinitely in favor of the environmental hunters, and only now led to the present global crisis, political disorder, and disorganization of both the human and animal gene pool, for the want of "female" intelligence. Most scholarship remains unable to see this because of a preference for males as sources of history. G. I. Gurdjieff explains why this could be so: the human essence class cannot be further evolved by Nature—that is, by natural means or *selection*—humans have come to a phase in their history where they must evolve themselves, based upon the form great nature has given them as three-brained beings and not by changing their DNA or augmenting their bodies with technologies.

> The evolution of man can be taken as the development in him of those powers and possibilities which never develop by themselves, that is, mechanically. Only this kind of development, only this kind of growth, marks the real evolution of man. There is, and there can be, no other kind of evolution whatever. . . .
>
> In speaking of evolution it is necessary to understand from the outset that no mechanical evolution is possible. The evolution of man is the evolution of his consciousness. *And "consciousness" cannot evolve unconsciously.* The evolution of man is the evolution of his will, and "will" cannot evolve involuntarily. The evolution of man is the evolution of his power of doing, and "doing" cannot be the result of things which "happen."[1]

As stated in part 1, the form of time surrounding the Earth was the only type of astronomy open to the Stone Age. A holistic phenomenon, centered on the Earth, required the Stone Age to build megalithic monuments to articulate it. But in recent centuries the collective has not been able to accept this simple truth behind such monuments: that they required the female form of tribal organization and intelligence, which industrialized cultures have suppressed in pursuit of growth through technology. The patriarchal Greeks, in fact, got many of their main symbols from the past application of female intelligence. And religion has now waned in the West, in favor of a pick-and-mix buffet of Eastern and indigenous practices and symbols, touted to come from a common origin, yet (just as soon as the reach back of ancient history runs out) up pops this question of a prehistory that cannot be understood, which is because it was not a patriarchal prehistory.

Cut off from the sky as a source of meaning and driven by (as we say) man-made environments, disasters, and wars, male governance is found wanting in the task of integrating/applying the female and male forms of intelligence. Again, as Gurdjieff had it, natural selection probably cannot provide the integration thing that he called the reconciling force. In potential, nature has already supplied both sides of the brain to both genders. And if humanity falls back into a new Stone Age, women would likely make a comeback to rule tribes once more with an affiliated work force of men, but humanity would then be set back again, by millennia, from the possibilities of human individuation.*

GENDER TRANSFORMATIONAL THEMES

One can recognize numerous themes taking the form of before, during, and after that appear to define the transformation of matriarchal Greece into our familiar classical Greece. The classical world, now seen as the beginning of Western history, was admired by the tenth-century Islamic philosophers and the elites of eleventh- and twelfth-century medieval Christianity in Europe, both highly patrilineal cultures. The touchstone themes listed below are indicative, in a general way, of the change in the cultural mind involved in this transition that began at the end of the Bronze Age.

---

*From Wikipedia: Individuation is a process of transformation whereby the personal and collective unconscious are brought into consciousness (e.g., by means of dreams, active imagination, or free association) to be assimilated into the whole personality. It is a completely natural process necessary for the integration of the psyche.

TRANSFORMATION THEMES

Minoan Crete	Archaic Greek	Classical Greek
Matriarchy	Heroes	Patriarchy
Orality	Epics and Myths	Literacy
Poetry	Rhapsody	Philosophy
Gods	Mores	Laws
Forms	Artifacts	Atoms
Metamorphosis	Drama	Causation
Saturnian Calendars	Festivals & Games	Solar Calendars

Taking an example from this table, the usage of heroes was different within the rulership of matriarchy and patriarchy. Formerly, Greek heroes could marry a "queen bee" and become a transient "sacred king," a consort for a year of 364 days and one additional day (making 365). This "year and a day" for the king's titular reign slowly changed as patriarchy increased its sway into our present usage: that the social status of heroes will lead to wealth and public office. It is quite releasing to see these missing aspects of patriarchal histories; it reveals that Greek myths were intended as a more creative form of historical record keeping than more modern histories, "written by the victors." The patriarchal authors and editors of the myths were illuminating the earlier norms and the key events in their displacement of matriarchy. These, through their own eyes, were power structures and sensibilities from a true though now quite obscure history, written only for that time but still influential today.

If the constellations were based on a rudimentary star chart from around 2800 BCE (see "Stellar Astronomy at Malta" on p. 31), then the Greek myths are likely a rewrite and palimpsest of an oral matriarchal set of tales. However intended, the achievements of matriarchy could not have been guessed at without these myths remaining somewhat true to the realities involved. The island of Malta gives us a deeper angle on these myths, that matriarchal astronomers became a primary inspiration for the religion of the matrilineal Minoans of Crete and thence the myths of later patriarchal Greeks.

ORIGINS OF THE MYSTERIES

If living in cities made citizens more functional, the cultural memory that numbers could be more than functional was lost, creating many outwardly quaint and implausible mysteries and turning sacred numbers into occult symbols. Pythagorean cosmology, involving numbers and musical harmony in the creation of the world, is perhaps the best-known example.

Using horizon astronomy, prehistory could only hope to measure average time periods between repeated celestial events. In contrast, natural science and physics has developed instrumentalities, such as the degree circle and telescope, to directly measure angles in the sky without using horizon astronomy and its limits to gain data only at those limited moments when the Sun or Moon rose or set. This has made modern science blind to how average periodicities* uniquely express significant patterns such as the Fibonacci golden mean ratios or the musical intervals between celestial periods. In a science where physical laws alone now define all outcomes, it is not expected that high levels of order should exist based on mere numerical invariances belonging to the whole numbers in their natural order {1, 2, 3, 4, 5, 6, 7, . . .}. The Pythagorean, Ptolemaic, and Platonic traditions preserved parts of the invariances found by the megalithic but have been unable to properly realize, or have lost the fact, that average celestial periods were the concrete origin of their own mysteries.

Mysteries inevitably arise when the direct factual basis has been lost, often due to technical, economic, and intellectual progress. For example, the number 12 had an exalted status in the twelve tribes of many ancient cultures,[2] the twelve signs of the zodiac, the twelve semitones in an octave, and so on. The ancient world found the first twelve numbers to be packed with harmonic, organizational, and constructional significance.

Jupiter-Zeus, as king of the gods in Greek myth, is directly linked to the zodiac and the Moon. In an average of 19×19 (or 361) days, Jupiter traverses one-twelfth of the zodiac, which is a single sign, and he stands opposed to the Sun every 398.88 days, which is 9/8 of the lunar year of 354.367 days. This giant planet is one-sixth the mass of the Sun,† but the next giant planet (Saturn) is twice as far away, smaller, and appears to have ruled the matriarchal worldview

---

*The direct pointing to objects and invention of clocks and orbital formulae have led to a dynamic astronomy in which an object's position can be calculated or predicted for almost any epoch or date.

†Jupiter, the nearest of the giant planets, is nearly the mass of a brown dwarf star.

FIGURE 4.1. Botticelli's *Birth of Venus* shows Venus-Aphrodite standing on the sickle-shell used by Saturn to cut off the sky god's generative parts, which were then thrown across the sky and landed in the sea. Rather like the rib of Adam, she is "Born of foam," and she governs the form of harmony based on the golden mean via the Fibonacci numbers {5, 8, 13}.

of Crete, preventing Jupiter from deposing Saturnian time by having Saturn (Cronos) "swallow" him at birth.

In a seminal story of the ninth century BCE, Hesiod said the sky had originally ruled by itself, perhaps meaning in a time before humans tried to understand it or quantify it. This rulership was by Ouranos,* or Father Sky, born of Earth Mother Gaia, who was his natural wife (see figure 2.5 and poem after). Saturn, born of these primordial parents, rose up against his father, and, having castrated the sky god with his sickle-shell, he threw his fathers "generative parts" (or creative power) into the ocean, to become Venus/Aphrodite, an inner planet brighter than Zeus when near to Earth (see fig. 4.1).

Cronos-Saturn had therefore created a goddess to carry forth the creation.

---

*The planet Uranus was named after Ouranos—quite appropriately since Uranus is a subliminal boundary beyond what the naked eye can observe and Saturn is the properly visible outer boundary of our planetary system. The planet is also, like Jupiter and Saturn, harmonic with the lunar year as 25/24, which is the chromatic semitone.

Venus becomes the brightest planet, like Jupiter but tied to the Sun, first dominating the sky as the evening star. Then she rushes between Earth and Sun (due to her faster inner orbit) and reappears as the morning star, as noted previously. The golden mean, or Φ (phi), and other invariances deriving from the number five, are primarily associated with living structures, especially in the pattern of their growth. It is hard not to compare her with Eve, the "mother of all living" in the book of Genesis (or the Virgin Mary) and of how she is symbolic of the matriarchal astronomy, and may have given Adam the fruit of the Tree of the Knowledge of Good and Evil at the center of the garden, perhaps the North Pole of the Earth. Saturn's act of castration can be seen as similar to the delineation of measurement vital to astronomical counting of time. The gender of Saturn-Cronos could have been changed since Cronos lived on Crete and was deposed by Zeus, who was clearly patriarchal and may represent a sacred king who escaped death on Crete.

The Moon is the third aspect of the goddess, a hag, after Earth itself and the young goddess, Venus, these forming the traditional triple goddess[†] of the goddess culture. The Moon seems to have tilted the Earth (in our scientific creation story) and created all manner of other necessary conditions for life on Earth, such as coastal regions and tectonic plates. Saturn has a synod with the Sun that is 16/15 (a semitone) to the lunar year, but, in contrast, Venus is directly linked to the Sun and the Earth itself, through the golden mean. (We will come back to these Fibonacci relationships in chapter 5, on Crete.)

Saturn-Cronos is later called Father Time (Chronos[3]) because he sets the cosmos into motion by castrating the eternal sameness[‡] of the rotating starry sky and creates a concrete (lunar) calendar beyond night and day, where the sidereal lunar months of 27 to 28 days became the primary markers in a society preceding the first-millennium patriarchal tribes of Greece, a calendar based on 52 weeks and 13 such months, the Saturn synod being 54 weeks long. Having sacred male kings, matriarchal societies respected the contribution of men but prevented their rulership from being absolute by replacing the male king every Saturnian year of 364 days, after which they were sacrificed and another chosen after "a year and a day." For this reason, Hesiod's story characterized Cronos as eating his own children, because one of them was fated to overthrow him just as

---

[*]The solar year being another type of conjunction of the Earth with one of the four pillars of the year, the two equinoxes of spring and autumn and the solstices of winter and summer.

[†]The three forms are maiden, mother, and hag or crone.

[‡]Plato's word for the starry sky, ever rotating but forever the same.

FIGURE 4.2. This object is called an "incense burner" perhaps because of its many holes. In fact, it was a calendrical object from which a calendar for eclipses can be deduced by counting days and the Saturn synod's division of the lunar year into 15 units (central circle of holes). The 38 eclipse seasons are the ring of 38 holes, inset from the outer ring.

he had overthrown his father. And who could be the brighter and more powerful than Zeus-Jupiter.

I re-created a lost calendar for Saturn from a painted "incense burner" found in Knossos, probably made by the matriarchal culture of the Minoans.* This counting device exploited the 16/15 relationship of Saturn to the lunar year to count the 38 eclipse seasons, found within the Saros cycle of repeated similar eclipses every 18 years and 10 days (223 lunar months; see fig. 4.2). The famous Phaistos disc achieved comparable results by counting lunar months directly. Chapter 5 has more on Saturn's matriarchal astronomy and calendars.

The motif of the lunar eclipse and the substitution of a stone for the child Zeus appears to link Saturn with this highly evolved usage of counting lunar

---

*This was noted by me on a visit to the Heraklion Museum and was eventually written up in previous books and websites. Full details can be found in the appendix on calendrical objects, which also details the Phaistos disc and the use of patterned rock, and so forth.

FIGURE 4.3. *Left,* Metis under Zeus's throne. © Marie-Lan Nguyen / Wikimedia Commons. *Right,* the birth of Athena from Zeus's head after he had swallowed Metis, showing Hephaistos cutting open the head of Zeus to release her, reborn as the goddess Athena, in a parody of natural childbirth. About 575–525 BCE. London, British Museum B 424 © British Museum.

months (stones) so as to predict celestial events* as opposed to counting the time in days between celestial events. Cronos's wife, Rhea, hid Zeus rather than allowing him to be swallowed by his father, who was instead presented with a stone to swallow, in swaddling clothes. Gaia-Earth then made sure Cronos could not hear baby Zeus crying (she suspended Zeus between the ceiling and floor of a cave), implying Cronos and Zeus were both in Crete at that time. Zeus subsequently defeats his father, and his rulership by patriarchy is then clear from the Greek myths, in which Jupiter repeatedly creates new children through any means possible, including through a virgin and through rape, and the goddess Athena is given a patriarchal birth through Zeus's head, as the transformed mother goddess Metis, his first wife, whom he swallowed! (See fig. 4.3.)

There was therefore a period in Minoan Crete when the goddesses and their matriarchy were culturally distinct from the ancient Near East, the Levant and Egypt. Saturnian and eclipse astronomy provided a calendrical solution to time through a 364-day year (resonating with Saturn) "and a day," making 365 days a practical year in tune with Venus, in Fibonacci fashion. Two millennia after that, we find Cappadocian Christianity using the matriarchal metrology and astronomy while, more than eight millennia before, Göbekli Tepe (in modern-day southern Turkey) used the same subunits of 12/77 feet as an enigmatic precursor to the Maltese megalithism inherited by Minoan Crete.

---

*Perhaps meaning lunar eclipses, when the Moon is swallowed by the Earth's shadow.

5

MATRIARCHAL CRETE IN
THE BRONZE AGE

As the megalithic age in Malta was concluding, the Minoan civilization in Crete was beginning. Crete is the fifth largest island in the eastern Mediterranean, and in its central northern region, inland of the port and capital, Heraklion, was the largest palace of the Minoans, Knossos.

THE MONOLITH BASEMENT AT KNOSSOS

In an early part* of the new palace, one can see a remarkable rectangular room with a central pair of square pillars, facing one another. Each pillar is 7 by 11 Göbekli subunits of 12/77 feet; that is, one Sumerian foot of 12/11 feet by one Royal cubit of 12/7 feet. Between the inner faces of the monoliths are 33 units of 12/77 feet, therefore equaling 3 Royal cubits. This repeats the motif, found at Göbekli, of two centrally placed and facing pillars—here a whole and significant number of subunits, related to the Sumerian foot of 7 and Royal cubit of 11 subunits.

The surrounding floor is a walled rectangle whose east–west dimension is 4 by north–south 3, with a door. It is therefore a rectangle formed of two counterflowing {3, 4, 5} triangles connected by their diagonals of length 5. The units of these triangles are 27 subunits of 12/77 feet so that the side length 3 becomes 81 subunits, and the 4-side 108 subunits. Remarkably, the perimeter of the rectangle is 2 × 108 (216) plus 2 × 81 (162), making a total of 378 subunits, the number of days in the synod of Saturn (see fig. 5.1).

---

*From the period called Early Minoan 1a (3500–2900 BCE), which is the start of the Maltese temple-building period.

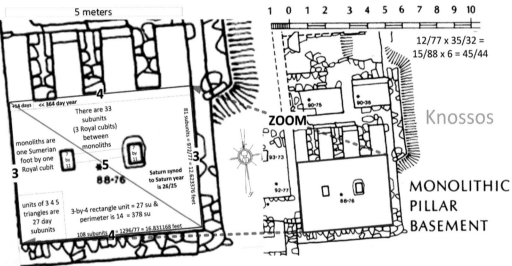

FIGURE 5.1. The monolithic pillar basement, rational in units of 12/77 feet. *Above,* after excavation. *Left,* rectangular room plus ancillary chambers and door top left. *Right,* the whole basement.

The top left passageway is 14 units wide, so that the perimeter between its western and eastern jambs (traveling counterclockwise) is 364 subunits; that is, the Saturnian *year,* in subunits per day. The 378-day synod and 364-day year, related by their difference of 14 days, are 26 to 25 fortnights long, respectively.* This references the matriarchal calendrical system of Cronos,† the Saturnian

---

*The Saturnian synod and year coincide every 25 synods of 378 days and 26 Saturnian years of 364 days.

†Cronos as Saturn is the god-smith of time, or Chronos.

god described by Hesiod as evoking a new world by castrating Ouranos, the sky god. Cronos probably represented the cultural astronomy of the goddess, who tries to suppress the arising of male kings (her children), who are liable to usurp her. Her last son was Zeus-Jupiter, who, as her successor, established the patriarchal pantheon of twelve gods, associated with the Indo-European language of the patriarchal Neolithic. Jupiter controls the Moon through his synod's 9/8 ratio to the lunar year (398.88 days). He also displaces the 27–28 lunar mansions of her orbit with his own transit of 1/12 of the ecliptic every 361 (or 19^2) days. Hesiod says Saturn is relegated to a golden cave on a smaller synodic island (378 days) of 16/15 of the lunar year* to which Cronos was exiled by Zeus, without spilling more divine blood, as with Ouranos, and hence invoking the Furies.[1]

In the matriarchal system, a queen-bee goddess would marry a hero-king who only had power for "a year and a day," after which time he would be killed.[2] It is therefore likely this chamber was at least symbolic of this calendar if not counted in practice as a calendar. The similarities of this chamber to Göbekli Tepe's enclosures in the facing pillars (or portal) and the employment of the same subunits, as also within Maltese megaliths, is striking. And the synthetic Saturnian "year" of 364 days, one short of the 365th day, strengthens the link between megalithic Malta and Bronze Age Crete found in Greek myths.

It is quite unexpected that two {3, 4, 5} triangles, with a unit of 27 subunit-days, when arranged as a 3-by-4 rectangle, lead to a rectangular perimeter equal to the Saturn synod of 378 in days. But when one looks at the 5-side of such triangles, their length of 135 subunits is the synod of Jupiter in tenths of a lunar month, truly swallowed by the mouth of the rectangle like the jaguar mouth of the Olmec of Mexico. This could show how Hesiod's Cronos would go on swallowing potential heirs, until Zeus was saved by his mother, Rhea, with support from the earth goddess, Gaia. A suitable scenario would be that Rhea, Gaia, and Cronos were all female. Zeus was Rhea's child as per the story, the female queen bee Cronos marries Zeus; Zeus lives on beyond his sacred reign, and the swaddled stone was then an unfortunate proxy who died, after "a year and one day," instead of Zeus himself; this might demonstrate how myth hid actual facts within a similar but suitably mystifying form. Graves says the limitations put

---

*The synod of 378 days, divided by 16 subunits and multiplied by 15 gives 354.375 days, 12 minutes longer than 354.367 days, the lunar and solar years not dividing by the seven-day week, which then seems Saturnian.

on a king's reign broke down in just this way: first the period of reign became extended to 60 lunar months, a proxy being killed to fulfill the ritual requirement. A longer period was next introduced, followed by no time limit, perhaps after a natural disaster such as an earthquake or eruption, Crete sitting on its own tectonic plate between Europe and Africa. If so, this explained how the matriarchal world on Crete came to an end.

The Saturnian calendar of 364 days, or notional 52 weeks in 365 days, is still active as the number of seven-day weeks in the year, which was an advanced solar calendar with additional but rough correspondences to the lunar orbit of 28 days, rather than the normal lunar year of twelve times 29.53 days. The synchronicity of the seven-day week to the synod of Saturn, of 54 weeks (378 days), and the Jupiter synod, of 57 weeks (399 days), aligned the Saturnian year, of 52 weeks (364 days), to the musical harmony between the lunar year and these two outer planets, which is the domain of Jupiter-Zeus. This makes the monolithic basement at Knossos highly relevant to the unique Minoan approach to astronomical time. The goddess was centered around Saturn, who created Venus, as per the myth of Cronos,* and linked to Zeus-Jupiter through the role of men and kings in goddess cultures.

As already stated, pre-numerate matriarchal societies used methods based on number notation, geometry, and counting processes unfamiliar to our calculational mathematics. The seven-day week is another trace element of a matriarchal and megalithic way of solving problems, the Sumerians and Hebrews revealing, through their own shared story of creation in seven days, a pre-patriarchal pattern for time. Matriarchy lurks in the patriarchal subconscious because humanity was Mesolithic before becoming Neolithic, with the change to the latter taking thousands of years to become widespread.

It must have been found that a 3-by-4 rectangle, whose perimeter is 14, would have a perimeter equal to the number of days in the Saturn synod if the unit size was 27, since 27 × 14 is 378. The floor of the basement may have been a symbolic version of a larger count, or at least a version in which days could be practically counted using distinct cups or holes along the perimeter. After this,

---

*For instance, the dynastic Egyptians had a similar slipping calendar of 360 days alongside the Sothic calendar of 365.25 days. From this, one can calculate the whole of time in a cycle of years that would return to its original starting point after about seventy years. There is some archaeological evidence, such as a distinctive Cretan fresco, that suggests the pharaonic line corresponded with the Cretan matriarchy.

5 × 27 gave the Jupiter synod of 13.5 months, in tenths of a lunar month, as 135. This gives Jupiter the musical interval to the lunar year of 120 tenths of a month; that is, as 9/8: the Pythagorean tone and spine of any tuning theory based on octaves and musical scales.

THE KERNOS OF MALIA PALACE

The 33 subunits between the twin pillars of the Knossos basement is a number also found in the coastal palace of Malia, directly east of Knossos. There are 33 cups in a ring on a large flat circular stone (a "kernos") adjoining the central court (see fig. 5.2). Central courts were an added feature of the "new" palaces, introduced after the old palaces were ruined by one or more disasters, around 1700 BCE.

This kernos would more normally be made of pottery, but at Malia it is a circular slab of stone with 33 small and evenly spaced cup marks with a gap left for a much larger 34th cup. This kernos almost certainly presented the fact that 33 mean solar months (MSM)* equal 34 lunar months, so that the excesses of the mean solar month over the lunar year (0.90625 day) become a whole lunar month after 33 months. This also means that, multiplied by 12, 34 lunar *years* will foretell 33 solar *years,* significant to the ancient world as the period of the solar hero (see fig. 5.2). In worldwide mythology a solar hero dies after 33 years.

The kernos is located directly east (currently az. 91 degrees) of the monolith basement, by 18.613 miles of 5280 feet, using Google Earth, whose imagery can, on a good day, resolve both the pillars and the kernos. The number 18.618 is another symptom of the remarkable system of time on Earth, interlinking the day, the lunar nodes, eclipse year, solar year, and the larger periods of 33 solar years and nodal period of 18.618 years. The number 18.618 is the smallest number, through the equal area geometry (and the actual value of π), to create, as radius, a circle of the same area as a square of side length 33: the equal area geometry of circle and square.

How strange is it, then, that the distance of 18.618 miles should separate the Malian kernos of 33 cups from the monolith basement of Knossos? Surely a mile was being used as a year.

---

*These are solar months, made of equal lengths as 1/12th of a solar year.

The Knossos Monolith Basement

© Steffen Löwe CC BY 3.0

The Malia Kournos

18.618 miles of 5280 feet

GOOGLE Earth

FIGURE 5.2. The kernos by the central court of the new palace of Malia has 33 small cup marks and a larger 34th cup mark. It survived the tsunami of Santorini presumably because it was built into the floor—a position that indicates its primary function was for court activities. It is placed exactly 18.618 miles from the Knossos basement, which *in years* would be the 6800-day period of the Moon's nodal period. The Moon's two nodes (crossings with the path of the Sun) cause eclipses either the Sun or the Moon when either body crosses the nodes. Each node takes 18.618 years to go around the ecliptic, in the reverse direction to the Sun and planets. If one divides 33 years (12053 days) by 6800 days, the result is 709/400 due to a common factor of 17 (see box below).

The Solar Hero of 33 versus the Dragon of 18.618

The number 18.618 is

- the number of days in which the two eclipse nodes of the lunar orbit move one day in angle, an angle due to the Sun's motion in one solar day: 18.618 days is therefore a node day;
- the number of years the nodes each take to traverse the whole ecliptic, called the nodal period;
- the square root of the eclipse year of 346.62 days, as the time taken for a lunar node to meet again with the sun; and
- the number of days between an eclipse year and a solar year, so that the solar year, minus the eclipse year, leaves 18.6222 days, just 6 minutes longer than 18.618 days.

These facts of time on Earth all derive from the motion of the nodes, in one solar day of 1/18.618 of the Sun's motion, along the ecliptic, in a single solar day.

And as stated above, 18.618 is the number of years that, multiplied by √π, equals 33 years, which leads to the equal-area model where a circle with radius 18.618 has the same area as a square with side length 33. This means 33 is the *smallest* whole number for which *any* real number can evoke equal areas between a circle and a square.

The 33 to 18.618 year relationship, true astronomically from Earth as the number of years of the nodal and solar hero periods, is implied in our iconography of the dragon slayers such as Saint Michael, where the lance that kills the dragon is the straight path of light against darkness, symbolized by the eclipses taking place at the two nodes of the lunar orbit (see fig. 5.3).

The numerical reason for 33 years is found in the approximation of the fractional part of the solar year of 365.2422 days being nearly 365.24, whose fractional part is then 32/132 days: 132 is 4 × 33 so that, after 33 years, 32/4 = 8 extra whole days result. The 365 days of a year times 33 equals 12,045 days; plus 8 days equaling the 12,053 days in 33 solar years—a whole number of days as is the case with the nodal period of exactly 6800 whole days.

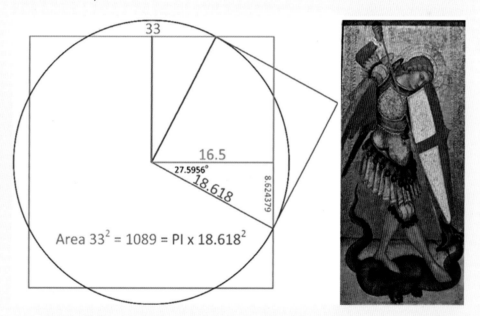

FIGURE 5.3. *Left,* the equal-area model of the nodal cycle, as a circle with radius of 18.618 (years) and the square with side length 33 (years), geometrically represents the solar hero, *right,* who slays the restrictive dragon with a lance, a polar symbol.

The makeup of 12,053 in terms of factors is of interest, too, as 12,053 is 17 × 709, a large prime, while the 6800-day length of the nodal period is 17 × 400 so that the two periods are related as 709/400, by removing 17 as a common factor, and 709/400 equals $\sqrt{\pi}$ to 1 in 3840 (using the actual, irrational π)! This removal of 17 was exploited in the dimensionality of the equilateral triangle relating enclosures B, C, and D (chapter 2) at Göbekli Tepe. In addition, the Preseli geometry of chapter 3 was a larger version of the same equilateral triangle, then using megalithic yards and standard yards in ratio to each other, of 32/29.

The 33-year period manifests as the Sun's rising (or setting) being in the exact same position on the horizon to the east when rising (or west when setting) when seen from a viewing platform, *at the same equinox after 33 years;* that is, one needs a good equinoctial alignment to detect the 33-year period. And the twin-pillar gap of 33 units at Knossos must, for all of these reasons, represent the solar hero in the context of the "year and a day" calendar of the basement room. The further reference to 33, of the kernos, built into the floor beside southeast corner of the main courtyard of Malia's palace complex, shows another 33, alongside a larger 34th cup (representing the Moon), relating to the fact that in 33 solar years exceeds 34 lunar years (of 12,048.478 rather than 12,053 days), some 4½ days less, useful in preparation for the end of 33-year observation.

THE HORNS OF CONSECRATION
AS ALIGNMENTS

It is likely the famous "horns of consecration" were a Bronze Age sighting tool defining an alignment, hence marking calendrical events. The Sun or the Moon are of similar enough size for either to display their full disk within such horns when they are rising at their extremes to the north and south of east, seen from a shrine palace or other site. These extremes would then relate to the nodal cycle of the Moon, lasting 18.618 years, and the solar-hero period of 33 years. In addition, the Moon and Sun fit the gender roles of male and female in the Minoan's Saturnine cult where the hero-king lives for a year of 364 days and a day, taking the 365 whole days of the Sun's path in a year. Men depicted bull-leaping and engaged in heroic sports that would later coalesce around the heroic stadia of Iron Age Greece, and the Olympian period between the games at Olympia on the Peloponnese peninsular, south-west of Athens.

FIGURE 5.4. *Above,* clay model of the "horns of consecration," words coined by Arthur Evans. Originally carved ca. 1500 BCE to 1450 BCE. Archaeological Museum of Heraklion. Photo: Zde CCSA 4.0. *Below,* a modern reconstruction based on stone fragments found at Knossos by Evans.

FIGURE 5.5. The use of "horns of consecration" to bracket the Sun and Moon at their extremes. *Left,* the rhyton of Kato Zagros, and *center,* detail tripartite shrine and courtyard featuring three altars. *Right,* the likely three-dimensional appearance of the whole sanctuary. See Shaw, "Evidence for the Minoan Tripartite Shrine."

At a specific viewing place, the horns would show the full disk of the Sun or Moon sitting momentarily between them. Seven points on the eastern horizon are of interest when following the nodal and heroic periods. For the Sun, the solstice extremes of winter and summer, and especially the equinoxes would allow one to see both the four-year and thirty-third year recurrence of the Sun at equinox, when the whole Sun fits exactly within the equinoctial horns. In the midpoint between its solstice standstills, the equinox is the best time to detect these recurrences, since the sunrise or sunset moves fastest on the horizon, day-to-day; other numbers of years, will not place the Sun exactly in the middle of the horns. Any alignment to the Sun will recur after any period of four, eight (99 months), or thirty-three years, and by moving your point of observation, the same horn can show a series of equinoctial recurrences.

If astronomy preceded religious ideas, then religious symbolism is highly likely to have roots in the processes, tools, or phenomena of astronomy, and, while temples seem a great place for "sacredness," many megalithic observatories have been called temples without any direct evidence; the assumed "sacredness" of "temples" prevents comprehension of the actual origins of a symbolic structure's form. Such a symbolic structure may have been created to do megalithic-style astronomy, where, in this case, one needs good far sights to known horizon events of the Sun and Moon.[3] It is highly likely that behind the "horns" lies the realization that horns can show the direction to something by accurately framing it. In an elevated and mountainous horizon, the peak sanctuaries of the first palace period found a new place in the new palaces, and their new central

FIGURE 5.6. A partial view of the "grandstand" fresco, fragments of which were found at Knossos and restored. It records a large meeting of young women around the central courtyard, probably that of Malia.

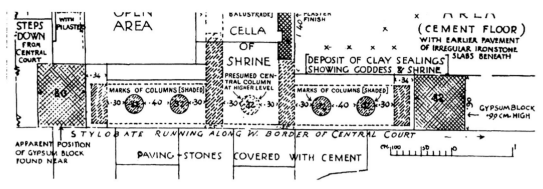

FIGURE 5.7. The tripartite shrine, or "loggia," at Knossos based on archaeological excavation.

courts show the further journey of such horns toward the more religious and public aspect. Figure 5.6 shows the pillars holding the triple frieze of five horns, with two pillars in the center and one either side, but Knossos had only one pillar in the central bay and two pillars either side (see fig. 5.7). This means that the frieze shown in figure 5.6 was most probably at the palace of Malia.

At Malia, with so many numbers of the lunar maximum, minimum, and thirty-third year equinoctial alignments possible, the distance of Malia from Knossos might have emerged from horns where, in the thirty-third Sun year, maximum Moon and minimum Moon would head to seven alignments: one east and the other three repeated to the north and south of east. In mountainous Crete, the directions of other palaces and mountain sanctuaries would also be important (see fig. 5.8).

FIGURE 5.8. The southwestern gate at Knossos. Alignment to other connected sites could also be marked, as with the porch (or portal) where Mount Juktas, the sanctuary of Knossos, is seen from the palace. Photo by author.

It may well be that the progress of the astronomical cult of the Minoans was the reason the sanctuaries became less significant and the religious life of the palaces more important.

Factorizing 18.618 Miles

The distance of 18.618 miles is already factorized, by Google Earth, into a real number measurement as well as the unit of measure—the familiar mile of 5280 feet. Surprisingly, though, a further division of the mile by 18.618 yields a familiar number of 283.6 feet, seen as the center-line diameter of the earli-

FIGURE 5.9. Plan of the Malia palace, whose area is revealed (in yellow) as being a factorization of 18.618 miles (98,303 feet) as the area 346.62 × 283.6 equaling 98,303 *square feet.*

est stone circle at Stonehenge, the Aubrey Circle. When 18.618 is multiplied by itself, the eclipse year of 346.62 days results, in day-feet. We can therefore see that 18.618 miles could have been arrived at and seen by the Minoans as 346.62 × 283.6 feet, if they chose Malia's kernos to be 98,303 feet east of the monoliths of Knossos. And this is supported when looking at the Malia palace plan in figure 5.9.

This new finding reveals a new factor (283.6 feet), which is 1/18.618 of the mile, just as the movement of the Moon's node per day is 1/18.618 of the movement of the Sun in a solar day (of 24 hours). In 18.618 days, the Moon's nodes move the same amount as the Sun moves in a single day. Because in 18.618 days, the lunar nodes move by one day of solar motion, the nodes travel around the ecliptic in 18.618 years. We already see the eclipse year as the square of 18.618 days (18.618 × 18.618 or 346.62 days long) and the solar year as 19.6177 × 18.618 days long, which is 18.618 days longer than the eclipse year.

If the mile is lengthened by 283.6 feet, then the result of 5563.6 feet is 19.618 × 5280 to represent the solar year. This only becomes clear if one sees the length 283.$\overline{3}$ (850/3) feet in inches as 3400 day-inches, which is *half the nodal period*. This length is found as the *inner* diameter of the Aubrey Circle of Stonehenge, while 283.6 is the *mean* diameter of that circle.* Therefore, the mean (283.6 feet) gives us the factorization of the 5280-foot mile into 18.618 parts, while the inner diameter of the Aubrey Circle gives the day-inch count for half the nodal period; that is, the day-inch count for half the nodal period is three inches or so shorter than a mile divided by 18.618. This corresponds with the fact that the eclipse year is twice the eclipse season of 176.31 days, the time taken for the Sun to go from one node to another.

This confluence of meaning, between the area of the palace and its distance from Knossos, implies how the factorization of real numbers was avoided by using rectangular areas such as the area of the building and known day counts. Another geometrical approach was the fact that 33/18.618 is the $\sqrt\pi$ of 1.7725, which makes the equal-area model work. But in this case, one notices that the diagram of equal areas has fourfold symmetry so that each quadrant is identical. This fact meant the geometry of the palace could form a square of side length 33/2 = 16.5, which multiplied by 18.618 gives 307.187 feet, so that two of the square's sides equal 33 units of 18.618 feet, as in figure 5.10.†

---

*Established in an official survey by Alexander Thom.
†In chapter 10, fig. 10.6, we see that the equal perimeter model was also shrunk by ¼ at St Peter's in Rome to overlay, on one quadrant, all four quadrants but with the Moon size unchanged.

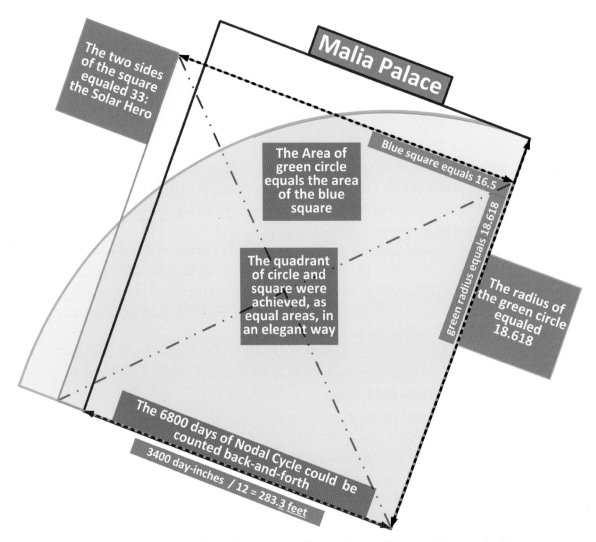

The two sides of the square equaled 33: the Solar Hero

Malia Palace

The Area of green circle equals the area of the blue square

Blue square equals 16.5

green radius equals 18.618

The quadrant of circle and square were achieved, as equal areas, in an elegant way

The radius of the green circle equaled 18.618

The 6800 days of Nodal Cycle could be counted back-and-forth

3400 day-inches / 12 = 283.3 feet

FIGURE 5.10. Malia palace as a single quadrant of the nodal period of 18.618 years equal to 18.618 × 18.618 multiplied by an implied 19.618 factor, like the factor 17 removed at Göbekli Tepe in figure 2.13.

And while the height of the rectangular palace was 346.62 feet (the eclipse year in days), the nodal period is 19.618 eclipse years, just as the solar year is 19.618 node days (of 18.618 days), leaving the solar day in total symmetry with the eclipse year. The time map in figure 5.11, suitable only for modern comprehension, shows the symmetry of time linking the nodal period to the solar day and the nodal period to the 33-year period in which the days within a single year, with a fractional part, resolves to 12,053 whole days. This emphasizes the

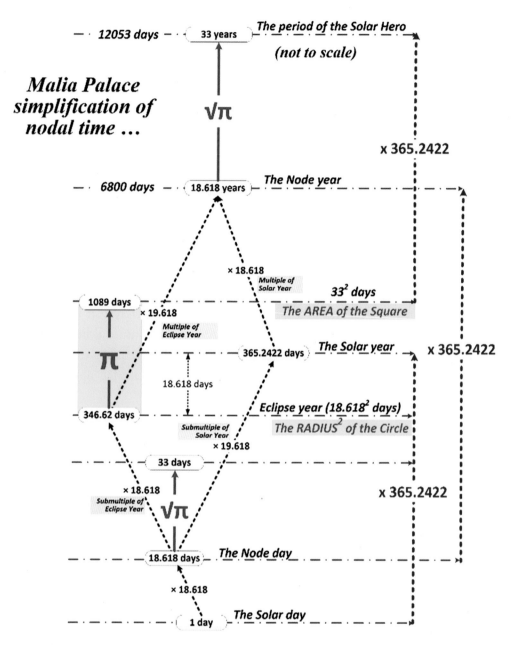

Malia Palace simplification of nodal time …

… since the Eclipse year is the square of the Node day (18.618 days), 1089 days (33^2) is π greater than the Eclipse year. Thirty-three days (or years) must then be $\sqrt{\pi}$ greater than 18.618 days (or years)

FIGURE 5.11. The time matrix using π, 33, and 18.618 to harmonize the day to the year and the nodal and eclipse periods of the Moon.

design-like nature of time in which the Earth and Moon have become locked within a perfect circle of recurrent time, 33 years long—requiring the special relationship that 33 and 18.618 have with π.

The Earth and Moon are two aspects of the triple goddess, and the sun was associated with a male sky god whose tears or semen impregnated the goddess with living things. The heroic nature of man's possible transcendence over fate gave rise to the Greek racetrack of a stadium of 660 feet, 1/8 of a mile, a complex conflation of time and space within metrology that also appeared, as if by design. The patriarchal hegemony of Zeus, born of Crete, retained much of this heroic bravado and gave rise to the Olympic Games occurring every four years, using the simpler leap year that, in India, was the Khumba Mela every twelve years.

THE OVAL HOUSE

A unique house was found east of Malia, a little palace or retreat belonging to the middle Minoan period MM1.[4] The house was not an oval but instead appears, like the enclosures of Göbekli Tepe, based on three equal circles, where the outer circles touch each other's centers and form an overlap upon which

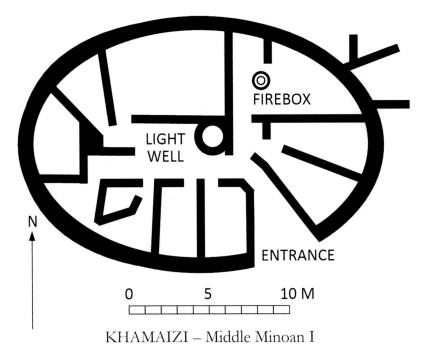

KHAMAIZI – Middle Minoan I

FIGURE 5.12. Plan and metrology of the Oval House in Khamaizi a vesica-themed house based on the cubit of 3/2.

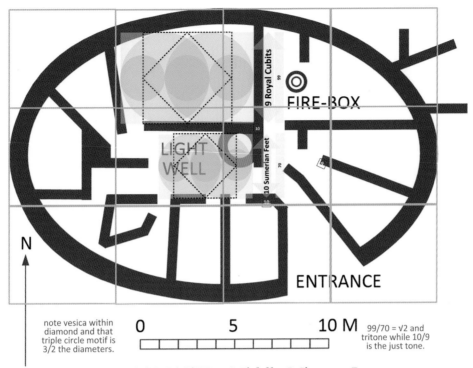

note vesica within diamond and that triple circle motif is 3/2 the diameters.

0   5   10 M

99/70 = √2 and tritone while 10/9 is the just tone.

KHAMAIZI – Middle Minoan I

FIGURE 5.13. Interpreting the Oval House and its main rooms. The house uses a 3:4 grid compressed to being 2:3 in proportion. The north room and well-lit room have ratio, in width, of 99/70, a rational √2, the north room being 99 to the well-lit room's 70 subunits. The length by width of the north room is another 3:2 and an oval within its walls is formed by three diameters to two diameters that form a vesica within the diamond of the dotted square. Oval shapes defined by triple circles were seen at Göbekli Tepe.

a central circle then enables an embracing outer wall, with an oval shape (see fig. 5.12). The rooms within seem small and utilitarian, and perhaps this was like the much larger palaces, and for the use of a matriarch. The house afforded a central hall with a light well, a large room north of that, and, by the entrance, a room with a firebox.

Using the scale given, this house seems designed using the 12/77-foot subunit. The oval shape derives from the nature of a vesica having two circles that have a combined length 3/2 of each circle's diameter, and this appears repeated in the rectangle of the hall and of the large room, but the hall is 10 Sumerian feet (70 subunits) wide and the large room 9 Royal cubits (99 subunits) wide.

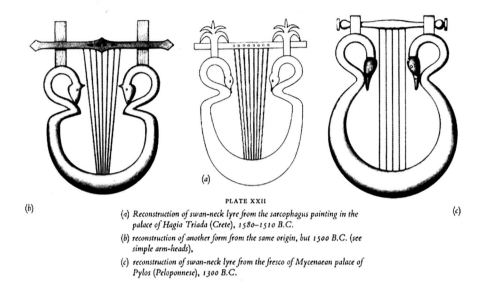

PLATE XXII

(a) *Reconstruction of swan-neck lyre from the sarcophagus painting in the palace of Hagia Triada (Crete), 1580–1510 B.C.*

(b) *reconstruction of another form from the same origin, but 1500 B.C. (see simple arm-heads),*

(c) *reconstruction of swan-neck lyre from the fresco of Mycenaean palace of Pylos (Peloponnese), 1300 B.C.*

FIGURE 5.14. Minoan swan-neck harps would have made harmonies in the sky audible through string lengths, a feat attributed to Pythagoras.

The numbers in subunits are in the ratio 99/70, which is the accurate ancient value for the square root of 2 (1.414).

The width of the hall can therefore become a side length that fits diagonally a square side length equal to the larger room's width($\sqrt{2}$). These design principles are influential for sacred art and can be seen to have sexual significance for a matriarchal culture. The house and main rooms are all organized within a 3-by-2 envelope (see fig. 5.13) within which the vesica geometry naturally fits. And given the musical intervals of 9/8 and 16/15 to the lunar year of Jupiter and Saturn, the 3/2 ratio can be seen as a perfect fifth, respected as divine by the later Pythagoras of Samos, and superior to the thirds of just intonation and modal music, which involve the human male number* 5 in their ratios of 5/4 and 6/5.

Rectangular Magic at the Oval House

A circle flattened in one dimension is an ellipse, and the Minoan seal rings were, like the oval house at Khamaizi, an oval approximation to an ellipse

---

*Male numbers were odd, as are the first two, 3 and 5, which are also prime numbers. Three was the divine male number in later Greek tuning theory and 5 the human male number in ancient tuning theory, as stated by Plato.

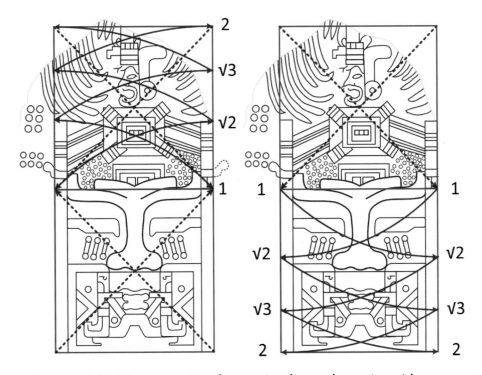

FIGURE 5.15. The properties of successive diagonals starting with a square naturally leads to doubling of the square into a rectangle with diagonal of √5, as in this overlay on the design of Tres Zapota Stela C, which is bounded by a 2 by 1 rectangle. See my *Harmonic Origins of the World.*

within a bounding rectangle of ratio 3 to 2, a cubit fraction. This ratio has an extraordinary relationship to the 4-by-3 rectangle, whose diagonal is a whole number, 5. When 4-by-3 is squashed within 3-by-2 by 8/9 (the whole tone), the harmonic model of Jupiter and the lunar year appears.

This 3/2 ratio was the size given to the plain of Atlantis of 3000 by 2000 stades* (see fig. 6.10), which is the ratio of the musical fifth, an interval existing between notes tuned in cycles of fifths, three pairs adding to the starting note, giving us our familiar seven notes† that fill the primitive octave, 2:1 of the starting note, to make a primitive scale, today's Dorian mode and

---

*A stade is 1/8 of a metrological mile of 5000 feet, and this length of 600 of any type of foot equals the various lengths of "stadia" racetrack lengths in Greece, either straight or around a rounded tailback. See Neal, *Ancient Metrology*, vol 1, book 6.

†The seven notes conveniently called out by choirs as {do, re, mi, fa, sol, la, si}, the solfège within the octave {do_1, do_2}.

ancient heptatonic. The octave rectangle of 2-by-1 (a 2-square rectangle) is the simplest integer rectangle, associated in ancient Egypt with the *djed* column, found bounding Egyptian iconographic art, and embodying the diagonal √5. The Olmec, in our Bronze Age, expressed this geometry in Stela C at Tres Zapotes (see fig. 5.15).

In surviving historical records, Pythagoras appears to be the first to realize that the octave was crucial to the notion of a numerical cosmogenesis, from the number one to higher numbers, but the ancient heptachord predated him as did just intonation, which perhaps goes back to the Sumerians and therefore Old Babylon, too. Geometrically, superparticular ratios are N:N + 1 right triangles, having sides one unit different in size. They were crucial geometry for ancient cosmology and its harmonic ideas that may even have taken the number 1 as monotheistic. Rectangles are an alternative manifestation of right triangles, because a rectangle of that sort is made up of two counterflow triangles, whose long sides form a shared diagonal and whose other sides are opposite each other.

We see that at Antioch, a mosaic visually presents a 4-by-3 rectangle (see fig. 5.16) using 4-by-3 circles, but the circles are squashed in the vertical, and the bounding rectangle, like the oval house, is actually 3 by 2. What follows in fig. 5.16 is a headache for the left brain.

A 4-by-3 rectangle can therefore be squashed to fit a 3-by-2 if the 3 side is squashed (to 3 times 8/9) while the 4 side is left as it is. By this method the Antioch mosaic has 12 rectangles that are 8 units high and 9 units wide instead of 12 squares. Remember that the height represents the lunar year, and this height is now 8/9 of 3. Since the Jupiter synod is 9/8 of the lunar year, then the squashed lunar year is now in the correct proportion to three of the original square lengths, as in fig. 5.17 and the oval house of 3/2.

In the oval house, the 4-unit side was instead conceived as having been enlarged by 9/8 rather than having shrunk the height. This suggests that the sacred king would enter where the synodic period of three lengths ends, at the entrance. He was Zeus-Jupiter as king for a Saturnian year of 364 days and a day.

If we use 4/77 feet, the length of the oval house is 1440 and Jupiter is 1080, and we have the full-sized harmonic model as the expansion of Adam as 1 + 40 + 4 = 45 in position notation, which then connects the house to the Parthenon, where the foot is 9/8 feet (the pygmy foot of Heracles of the twelve labors).

In the Parthenon (see fig. 6.7), the same rectangular magic reveals that Athens was using a foot 9/8 feet long to achieve the harmonic model

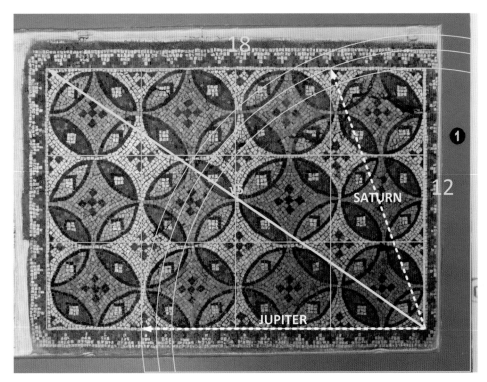

FIGURE 5.16. A Roman mosaic in Antioch of 4 × 3, or twelve, circles or months, but once enclosed in a 3-by-2 rectangle, the lunar year is shrunk vertically by 8/9, so that 3 squares on the horizontal are 9/8 of the lunar year. This also shows how the cosmic octave is 18 lunar months to 36, for two Jupiter synods to be 27 lunar months, 3/2 of 18.

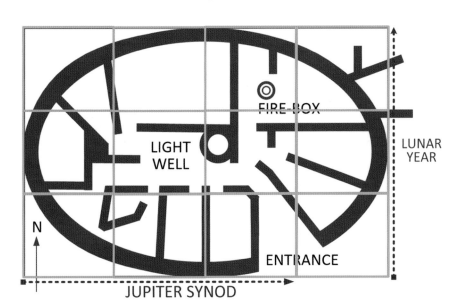

FIGURE 5.17. The oval house has a length of 480 and width of 320 made up of twelve 9-by-8 rectangles so that three 9-unit sides add up to 360 units of 12/77 feet—our thematic measure first encountered in Göbekli Tepe.

foundational for the Parthenon's walls; that is, a rectangular system for laying out monuments according to the harmonic model had become established in patriarchal Greece with the goddess tradition as the most likely source for that scheme. Zeus had his head cut open to reveal Athena: this reveals his head was the Earth, leading to her Parthenon who expressed her "mother" Metis from Zeus's head; that is, from a patriarchal perspective.

WHY MATRIARCHAL ASTRONOMY HAPPENED

It is quite impossible to recognize astronomical patterns within archaeological buildings without understanding the type of astronomy studied in prehistory—namely, the counting and study of average-time periods using lengths and geometrical models of circle and square, triangle and rectangle. By good fortune, I began to study the synodic periods of the Moon in 1993* and by 2000 had found similar number relations between the planets. During this study I needed to develop matrix diagrams, like that in figure 5.9, to grasp the numerical time patterns in as simple and intuitive a way as possible, as laid out in my first book, *Matrix of Creation*. Having unraveled the patterns, I moved to finding evidence that some of these patterns had been deduced in prehistory, since a lot of traditional information points to that. After twenty years, all megalithic activity appears to be rooted in the invariances of planetary time, and many monuments were often influenced by these invariances, not just in the units of measure but also in the presentation of geometries like the equal-area model of the 6800-day lunar nodal period and 33-year solar hero period; that is, the numerical astronomy of the megalithic was transmitted, by the megalithic, to the earliest civilizations as a template for religious storytelling and building design.

For example, long before Near and Far Eastern arithmetic, prehistoric astronomers could know how many days lay between Saturn's synodic loops, as the Earth overtakes Saturn on its inside "lane" or orbit of the Sun. This type of invariance in geocentric astronomy, and many others, are the same today as yesterday, meaning they are immune to the processes that destroy information. And once part of an invariant pattern is discovered, the rest can be reconstructed, just as Isis is said to have reconstructed Osiris after he had been cut into pieces by Set and his parts scattered along the Nile. We automatically

---

*Starting from Robin Heath's lunation triangle result in his *Sun, Moon and Stonehenge*.

think such stories were just made up, but they refer to a visibly eternal world of planets and what happens there.

Our modern story is different. Once it was rediscovered* that the Sun was the gravitational center of the planetary system, and since ancient astronomers were not thought able to resolve orbital periods around the Sun, it seemed that geocentric astronomy had little significance and so was largely discarded. However, it has turned out that the geocentric view from Earth, from where we observe, was more significant numerically than the idea of planets just accidentally going around the Sun. For a start, the Moon orbits the Earth as the result of a collision with the proto-Earth, and only later by the Sun to show the phases of the lunar month, making the Earth the actual center for the Moon's orbit. The lunar year then resonates musically with the outer planets, and the Earth resonates according to the golden mean with the inner planets, especially Venus.

The Stone Age mind would directly use the form of an astronomical problem to achieve a non-arithmetic solution to numerical problems. Unless we can find how their approach worked, we are barred from understanding what they understood. Celestial time periods seen from the Earth have something to say about how the environment of the Earth came into existence, but the process of how modern science brought humans back to astronomy has blocked our realization—that prehistory had discovered the world was not an accidental creation of material forces. This defines the significance of megalithic astronomy today, and it came from a different way of perceiving the world.

THE SIGNIFICANCE OF FORM
TO MEGALITHIC ASTRONOMY

It was natural to see the paths taken by the planets as roads in the sky, and the sphere of the starry sky as an analogue in heaven for the Earth, which had its own topography. Just as the Sun's road runs through the twelve zodiacal constellations (a "circle of animals"), this is really the motion of the Earth around the Sun, and the rotating Earth only seems static relative to a Sun only seeming to be mobile day-by-day and within daylight hours; in similar fashion, the stars as a whole move throughout the night. They are, in fact, effectively stationary but appear to move as a single whole, in the opposite sense to the Sun and

---

*Aristarchus of Samos was the first.

planets as the latter move along the Sun's path. Such mirroring behavior, the opposite of what it seems, is due to our view of the night sky from the planet of a star. And though we can see the point of view of the Sun in our mind's eye (as an abstract project of the mind), we do not live upon the surface of the Sun but rather experience the world from the surface of our planet.

Therefore, if one stops conceptualizing the night sky in a modern way, one can see the phenomena actually presented by the sky before scientific conceptualization and rationalization. Such suspended judgment can reward the viewer with insights such as that time has a form. And this was the experience of the ancient present moment, enlarged in scope by memories of sky events and the counting of time. New patterns emerged, according to natural principles, based upon a numerical grammar, expressing the experience of time for the mind, body, and soul of ancient naked-eye astronomers.

However large a present moment is, its center contains our purely subjective experience of time as the core experience of selfhood itself. Understanding the form of time expands the experiential present moment, and this, I believe, has been the crossroads for the two ways of ancient and modern thinking: in the past, naturally by form, and now intellectually by functional visualization. One can give a formulaic image to something while not understanding the essential form behind what the formula describes. The past cosmologies ask for the form to be understood, since they were found integral to cosmic consciousness (now a mere term) from which religion arose, where the world has a form that, to exist in time and according to physical laws, would *become* a modern formula governing a measurable fact such as orbital speed. The formula-based conceptualizations of planetary dynamics are inevitably blind to the form of these actualizations, based on whole numbers, geometry, and the average angular rates of bodies seen from the Earth. These seem to have been achieved by a higher intelligence as a form, and hence are religious, calibrated by the flow of time: the rotations of the Earth, the synods of the planets, and their integration with the Sun and Moon—and these could be understood using megalithic methods.

Statues Representing Knowledge

Astronomy, experienced from the surface of the Earth, is a naturally numerical system of time created, most probably, for the emergence of life and intelligence. Two completely different forms of harmony rule the geocentric astronomy of the Earth.

First, the Fibonacci series shaped proximate planetary relationships because orbits repeat, and they are naturally able, through their reduction of multiplication to addition, to approximate an integer golden mean that creates otherwise unlikely seeming coincidences within our astronomy of time. The golden mean is completely one of a kind, as the only ratio where its reciprocal and square can be achieved merely through adding or subtracting 1 to its value of 1.618034. This can reduce the complexity of planetary orbits, enabling intelligences (variously called the angelic host, demiurge, god-as-maker) to create special conditions for our living planet, perhaps a natural arrangement in suitable star systems for their inner planets near the "Goldilocks zone," with a large Moon, and so on.

The Cretan goddesses can be seen as the extraordinary result of thousands of years of geocentric astronomy involving women and men. The three forms of the goddess were widely projected on Venus, the Earth, and the Moon. Considering time on Earth as meaningful, the vision of a likely higher intelligence who made humans intelligent, evidently led to a religious explanation upon which Minoan society became organized, in part, around three goddesses. Two of these goddesses (Venus and Earth) are locked into a reciprocal golden mean relationship, in which in eight years of 365 days there are five Venus synods, as per the trilithon ellipse of Stonehenge discussed at the end of chapter 3. This can be seen once the "threatening" nature of the snake goddess is seen as personifying the consciousness of a geocentric astronomer.

The Minoans also appreciated that the outer planets were organized according to musical harmony relative to the Moon and its lunar year. Perhaps, since one dances to music, the lower part of the goddess represented outer-planetary musical harmony, as four layers of the flounced skirt, while the Fibonnaci harmony to Venus is shown in the in-body proportions of her torso (see fig. 5.22). Her snakes in either hand show the Sun's path along which the planets and Moon move on the ecliptic, at different rates, the tails of snakes being a standard replacement for the leg of celestial beings in the later Greek tradition, since they move along by a rippling of their "bodies" in time.

Interpretation of the Snake Goddess

The statuette in figure 5.22 represented the astronomy of the goddess astronomers; which related the zodiacal year, the Venus synod, and 945-day periods as two golden rectangles between her brow and the bottom of her dress. Above her navel shows her vision of the night sky, looking south, while below is an elliptical apron representing her area of astro-geometrical workings. The base of the figurine is the Venus synod, while the four layers above are the harmonic periods of the Moon, Saturn, and Jupiter, being a tone above and below 2 and 3 lunar years.

This, the smaller figure found, as restored, holds two snakes in her raised hands, and the figure on her headdress is a cat or a panther. As excavated, she lacked a head and the proper left arm was missing below the elbow. The head was re-created by Evans and one of his restorers. The crown was an incomplete fragment in the same pit, and the cat/panther (a good call based upon matriarchal hat of Athena) was another separate piece, which Evans only decided belonged to the figure sometime later, partly because there seemed to be matching fittings on the crown and cat. Recent scholars seem somewhat more ready to accept that the hat and cat belong together than that either or both belong to the rest of the figure. (Extracted from Wikipedia.)

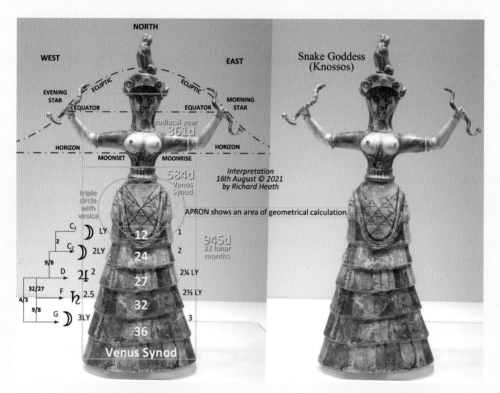

FIGURE 5.22. *Right,* one of two snake goddess figurines found at Knossos; *left,* my interpretation of the figure.

1. The Golden Mean: There are two golden rectangles in scale relationship to the golden mean. Hence, the shorter side of the larger rectangle has the same length as the longer side of the smaller rectangle, this side running through the navel of the goddess. Their common side length has three meanings:

The 584-day length of the Venus synod, which can the thought of as defining the unitary number 1 so that, the longer side-length below is of 945 days (32 lunar months), then the golden mean of 1.618; but it appears as a Fibonacci ratio of 144/89 = 1.617977.

In the same way, the smaller golden rectangle is actually the ratio 55/34 (= 1.61764) so that the zodiacal year of 361 days is the reciprocal of the shared side length (the Venus synod of 584, actually 583.92 days).

2. The Night Sky: Above the navel is the goddess and looking at what she is enacting with her arms (the horizon), her hands (the celestial equator), and the rising and descending snakes (the ecliptic). Her cap was later carried over to Athena's, when she pops out of Zeus's head with a cat upon it. The cap probably had 10 florets around its raised border (the number of Venus elongations in 8 years and 5 synods), a hat located in the celestial north, perhaps the circumpolar region with the cat as the North Pole (source of wisdom).

3. Hips and Geometrical Apron: The hips define a circle, and this can be drawn as two symmetrical circles, of the same size, which touch each other's centers to form a vesica, a matrilineal and Mediterranean theme that relates to the triple goddess and the cosmology of women as sources of creation. The vesica then occupies the apron's central area, and within it, a lozenge pattern of four sub-lozenges, each having two upper horizontal lines, out of a possible five. The sub-lozenge pattern is shown as continuing beyond the borders of the apron. Two 3-by-4 rectangles have been drawn in yellow within the central circle to show the usage of 3-4-5 triangles to count the Saturn synod at the Knossos monolith basement and the Metonic period (235 lunar months) at Ayvali Kelise in Cappadocia (see chapter 7).

4. The Dress: According to the two golden rectangles, the base of the skirt has a diameter equal to the Venus synod, the shared side of both golden rectangles. The dress is made up of 6 flounces, like conical sections, expanding downward from within the one above, each with alternating light/dark squares in cream and brown, with ten running around each layer, like the florets on her cap.

5. Its Musicality: The lowest layer is taken to be the Venus synod, and the highest (covered by the apron) could be the Moon as the lunar year of 12 lunar months. In between, the four remaining levels are taken to be 2 lunar years (24 lunar months) as the smallest and 3 lunar years (36 lunar months) as the largest.

Though there is no factual evidence for the notion of octaves and a "tonic" framework for music prior to the first millennium BCE, astronomy provides the invariance for it, and modern terms, based on the tonality of the octave, help the understanding. The span is 3/2, a perfect fifth, and the middle terms are thought to be 2 Jupiter synods (of 27 lunar months) and 2.5 Saturn synods of 32 lunar months, also 945 days (see fig. 5.23). Musically, 27/24 is a tone of 9/8 or a whole tone, as is also 32:36. Between 32 and 27 is the Pythagorean minor third of 32/27. This would form a subset of what modern musicology calls pentatonic tuning: 24:27:32:36 is that portion between the tonic and the fifth—lacking a fourth tone. The lunar year (12 lunar months) provides an octave drone to the 24 (2 lunar-year period) below it.

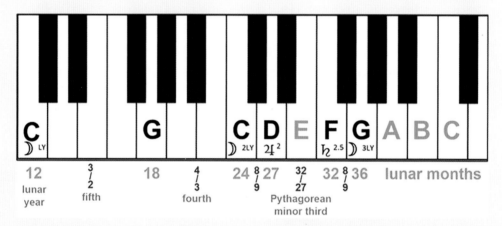

FIGURE 5.23. Modern keyboard showing where the proposed four notes of the goddess astronomer's dress sit, where the first four notes of our pentatonic scale would be in C-major's white-note octaves.

THE ICONOGRAPHY OF THE DOUBLE AXE

As a result of doing work on the geometry of the Cappadocian cross (in chapter 7), I became interested in the similarities with the sacred gold double axes of the Minoans that were similarly used to consecrate religious venues, especially their palaces. To explore this, I examined a famous example of a double axe, found in a sealed treasury of axes and other precious items at the palace of Knossos.

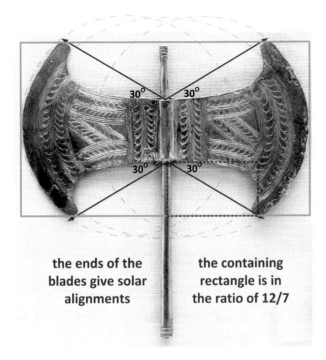

the ends of the blades give solar alignments

the containing rectangle is in the ratio of 12/7

FIGURE 5.24.
The geometry and alignments within the Knossos double axe.

I believe the axe was a symbol of the Minoan empire itself, taking the form of its geographical setting and extent with relation to the distinctive extremes of the Sun, at that latitude of 36 degrees north, of 30 degrees north and south of the equinoctial rising in the east and setting in the west, as shown in figure 5.24. One has to say that this interpretation may never be provable; in fact, because when something becomes obvious to oneself as a form, that perception is only implicitly obvious to others. The Minoans were, unlike us, centered in their perception of form rather than in functional description, in part because of the oral nature of their society. And if the Minoans saw their Mediterranean Sea empire as being at such a latitude where the Sun at the two solstices was plus or minus 30 degrees to east on rising and west at setting, then the axe blades relative to the handle sockets were probably made to carry this signature.

One can also see how this geometry was again evolved from two overlapping circles and a central circle, shown dashed in the figure. Also noteworthy is that this double axe is contained by a 12-by-7 rectangle, the ratio in feet of the Royal cubit, infamously Egyptian but, more importantly, one of the small set of measures we identified as useful at Göbekli Tepe, as eleven of the 12/77-foot units. The outer circles to left and right show how the shapes of the blades were arrived at through two overlapping and one central circle.

6

PYTHAGORAS, THE BIBLE, AND PLATO

There is a myth of higher knowledge being a subterranean river, in both India and Greece, which alludes to a sort of number symbolism within great time and has its origins in astronomy, gods, and goddesses. In the west of India, the river of knowledge is called the Saraswati, after the goddess associated with knowledge and music; this river disappeared underground as a result of the same desertification that afflicted the world after the Ice Age. The river Arcadia was said to run beneath the Peloponnese, home of the Olympic Games and of the martial Spartans, whose customs were close to matriarchal.*

The Peloponnesian peninsula is below the Gulf of Corinth and joined by an isthmus to Attica and Athens. After our Renaissance, Arcadia became a setting alluded to by poets, mystics, and secret groups, as if the river of that place were represented in the poetry of a geocentric sky and Earth, the poetics protecting the secret tradition: "Meters are the Cattle of the Gods."† That sort of Mediterranean tradition can be recognized through its methods and habits, being carried along in the river of time by those groups active within the tradition, through the megalithic and Bronze Ages, aboveground, each encounter refining and reforming the outer content, as new hands of either gender, engage with the tradition's necessary survival.

---

*A fuller list would be the ancient Greek city-states of Sparta, Corinth, Argos, and Elis, which all played a crucial role in Greek history and were prominent during the classical period.
†*Satapatha Brahmana*. See chapter 7 of Calasso's *Literature and the Gods*.

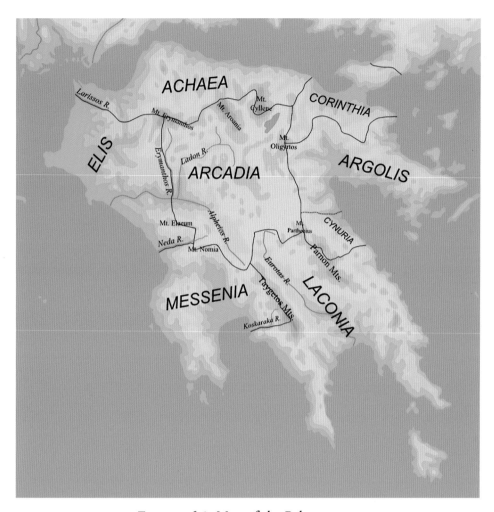

FIGURE 6.1. Map of the Peloponnese.

NUMBERS AS MULTIDIMENSIONAL SYMBOLS RATHER THAN LENGTHS

Ernest G. McClain* revealed the ancient tuning theory as being far from a primitive experimentation with the string lengths of harps or the fingering of flutes. There are magnificent tuning documents in cuneiform, including pictures of musical instruments upon seals, and tables of harmonic or "regular" numbers

---

*American Pythagorean and musicologist (1918–2014), author of *The Myth of Invariance* and *The Pythagorean Plato,* with whom I cut my teeth regarding ancient tuning theory.

up to 60^4, which equals 12,960,000. Such a table of whole numbers allowed the invariant structures responsible for music, whose origins lie within the invariant field of number, to be studied, when reciprocated, as musical ratios called "submultiples,"* which are exact divisors of a number;* that is, the denominators of all the regular numbers in all of the harmonic fractional ratios contain only the factors of 2, 3, and 5.

My question was: how had the lengths of astronomical time become translated, as numbers, into a well-developed numerical tuning system, and did harmonic number theory give birth to a type of theological storytelling concerning the gods of the sky and Earth? Certain sacred numbers have been dropped into ancient stories in a number of ways, such as ages, numbers of things, numeric values of names (as gematria, sequences, sums, or as exponents), numbers factorized by prime numbers, considered dimensions, and many others. McClain found that many of these textual references to number, when treated as the limiting number (high "do") for an octave, such as 360 to 720, contained a set of harmonic numbers between them, automatically forming the tones and intervals of different types of tuning systems. The most important of these sacred numbers then appeared to give a musicological, theological, and astronomical meaning to them. I then saw this technology as relevant to there being a musical creativity behind the planetary creation surrounding the Earth.

McClain called these number sets of harmonic potentials "holy mountains," and, within these, the symbolic life and theology of many religions seem to become evident, such as the Star of David, Noah's Ark, Indra's/Marduk's victory over Vritra/Tiamat. Multiple viewpoints from static traditions, to me, reveal a secret component of harmony in ancient texts supporting the notion that harmonic and astronomic parallelisms were a compositional technique associated with oral and scribal centers of ancient storytelling.

A length only becomes a number when we measure it as a number of units long. Megalithic metrology was all about using a range of units to make alternative measurements of the same length. To achieve the stage where numbers became symbolic or sacred, early numeracy had to exploit the numerical factors of lengths, requiring some form of metrology, and those transforms we know as geometry must have been used to transform numbers before arithmetic was

---

*Our rational fractions in the set {3/2, 4/3, 5/4, 6/5, 9/8, 10/9, 16/15} have been the most important to early music.

later developed. When numbers themselves became arithmetical symbols, this became the arithmetical way of working with numbers using arithmetic operations rather than the metrology and geometry of lengths; that is, when numbers became symbols, they took over from lengths as numbers.

Numbers became symbols in Sumeria around 3000 BCE, through a promissory technology, formed using clay markers in sealed clay envelopes, to account for the contractual transfer of goods. On the envelope was a symbol of the markers inside until people realized the number as a symbol of a set of things made the contents of the envelope somewhat redundant. Then, by the first millennium BCE, alphabetical writing would develop alphabetical letters-as-numbers, hence as symbols, by the allocation of a number to each of, say, 22 alphabetical letters: for example, alpha for 1, beta for 2, up to iota for 10, rho for 100, and omega for 800, in decimal. Many such gematriac systems exist for alphabetic languages, and in Greek and Hebrew words were chosen to express a sacred-number equivalent for a god's name or for the numbers of years someone lived, and so on.[1]

But secular arithmetic requires functional processes (such as addition and multiplication) to be performed on numbers-as-symbols, using a position notation first seen in the Sumerian sexagesimal base 60. Our decimal system is base 10, which removes the musical prime number 3 from the base system. Decimal number strictly imposes the position within a number to represent the powers of the base number. This had the benefit of limiting the *number of number symbols* to just ten or, at first, nine and a space: the zeroth power remained a space or later, zero-as-a-symbol was discovered in India as late as 600 CE. This gave us our secular numeracy—child's play when educated. But the move to this, as with literacy, took millennia, but numbers could still be lengths in architecture, gematria, and ancient texts.

Modern sacred geometry discounted number-as-length, perhaps because arithmetic was a different department of the traditional arts. In a 3-4-5 triangle, numbers-as-lengths are sublimated though the abstract numbers are often given in the center of each line. This shows us what number symbols are—an intensive magnitude rather than an extensive length. In modern math and especially topology, a number can be a node as the dual of a length: they form two versions of the same data, the *extensive* thing itself and the *intensive* number symbolizing it. A builder translates the number measured on a plan into a length, which is only sometimes notated with length.

The numbers within the base of the Sumerian sexagesimal number system were the prime numbers of the first Pythagorean triangle, the base 60 being the

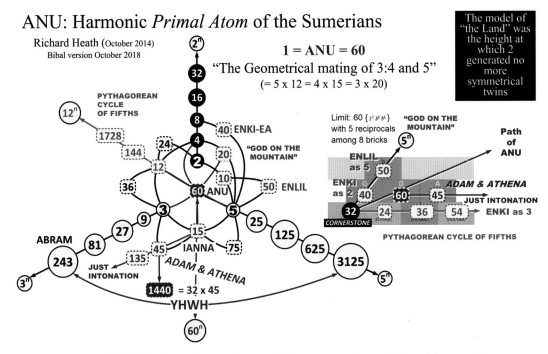

FIGURE 6.2. The Sumerian universe of Anu as a volume defined by numbers in the dimensions of their primes. Their small holy mountain for limit 60 (after McClain), is on the right, growing to the top right, where higher limiting numbers are created. Powers of five are seen as vertical and powers of three horizontal, with powers of two being added to place every number into the octave 30 to 60.

product of 3, 4, and 5. The primes {3, 4, 5}* gave the first three primes {2, 3, 5} the notion of three-dimensionality, as can be seen in my diagram (see fig. 6.2) of the three primordial gods of the Sumerians: Anu = 60, Enlil = 50, and Ea-Enki = 40, with the numbers from other myths also included.

Treating the first three prime numbers as dimensions to create a harmonic space naturally led to the secular concept of volume, and to this day "weights and measures" are an early version of numeracy useful for defining volumes of grain, areas of land, and so on. Archimedes's "a-ha" moment was when he saw the volume of his body displace the same volume of water, as with the myth that Pythagoras discovered numerical tuning theory using strings. There was an ancient interest in areas and volumes such as the "grain constant" of

---

*Four is two squared, the first square, and hence an area.

1,152,000 seeds of grain as a volumetric measure. This interest can be seen in the rectangular numbers (area) defined within the Great Pyramid to encode the degree distances between different latitudes (chapter 3).

Having sixty characters in base 60 required sixty symbols, and the decimal base 10, by dropping the prime number 3 within the base, left 2 as a simple doubling, and 5 × 2 as the fingers and thumbs of the hand, often used for impromptu counting. In similar fashion the alphabet provided decimal numbers-as-letters allowing names to be the sum of their letter-number values, and the possibilities of having equivalent numbers as sums and even products. For example, Adam has three Hebrew letters *A, D, M,* which are then 1, 4, 40 and when summed make 45, a powerful harmonic root, and through position notation are 1440, or 2 × 720. The number 720 is the limiting number required to form five modal scales when the numbers {2, 3, 5} are treated as three dimensions that, within a single octave, can be seen in two dimensions {3, 5} in an area of those harmonic products of {3, 5} (McClain's holy mountains) less than the limiting number: each of these products of 3 and 5 is then multiplied by two until the result lies within a single octave, in this example between 360 and 720. In figure 6.3, the tritones of Adam as octave limits {360, 720} or {720, 1440} are undeveloped, requiring further doubling to 2880. This implies Adam was limited to the range 45, his harmonic root, and 1440, his position notation, and was thus limited in scope until transformed into a creative being with access to the tritone, where the principle of creativity resides. This creativity of the tritones is what opposes the tonic (or key) who created the world, leading to the musical dilemma of Eden: to eat the fruit of knowledge or to remain a creature of circumstances not of one's own making.

The realm of harmonic numbers, being based on the first six numbers {1, 2, 3, 4, 5, 6}, and modal music, with multiple scales, requires only a few more early numbers {8, 9, 10, 15, 16} to make a low-number octave. Ignoring prime number 5 as a dimension, we become stuck with the single heptachord scale (our Dorian), based on {2, 3}, of seven notes that are less harmonious with respect to one another without the mediation of 5, hence requiring the death of the primordial but natural order that, in tuning theory, was characterized by Vritra in the Vedas and Tiamat in Sumeria.

These long snakes of tonality were, in practice, the heptachord achieved through a tuning method that is still the foundational "by ear." The Sumerians then evolved heptachords into seven modal scales, again "by ear." Later arithmetic techniques allowed music's strings-as-lengths to be manipulated using

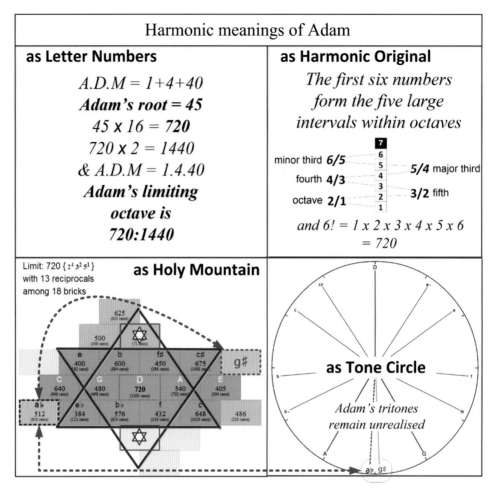

FIGURE 6.3. Four ways of seeing Adam as initiating the lost harmonic tradition within the Bible.

primes {2, 3, 5} as dimensions and the properties of the 3-4-5 triangle to map an octave's harmonic roots by noting how the sides belonged to the subset {1, 2, 3, 4, 5, 6, 8, 9, 10, 15, 16}! Figure 6.4 shows that {3, 4, 5} is the DNA of the ancient modal music of India and Greece.

In the use of bases, one sees a restriction in the number factors that will cleanly operate with respect to one another, which excludes larger prime numbers that must then be considered part of a measurement. The foot-based units used since the megalithic were similarly constrained to using the first three primes plus the next two primes, 7 and 11, since 22/7 was an accurate approximation to π. Therefore:

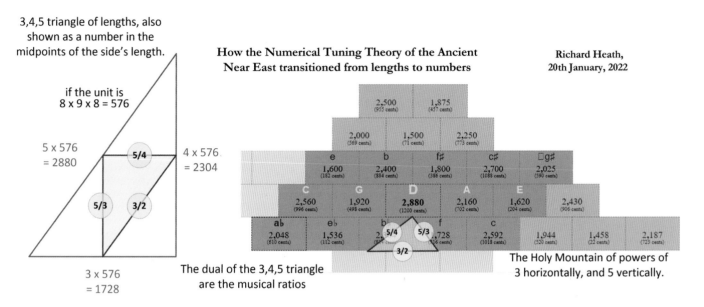

FIGURE 6.4. *Left,* placing numerical lengths at midpoints of a 3-4-5 triangle, the large intervals of modal music appear between the three sides. *Right,* a mountain of ascending powers of 5 made of horizontal rows of the ascending powers of 3, natural to octave tuning. The limiting number is 2880, and beyond Adam. This shows how dimensional separations of powers could reveal the musical possibilities within the number field as being a field of vector intervals, from each number to every other number. When translated into astronomical time relative to the lunar month, it was possible to see that one of the tritones was Saturn, who was also the bottom left cornerstone—a pure power of two whose five synods of 378 days equates to 64 lunar months (= 2 × 945 days).

- Early numeracy was about manipulating the earliest prime numbers as geometrical lengths.
- Bases, such as 10 and 60, were attempts to conveniently store numbers-as-symbols—rather than as lengths, counters, letters, or lists—of their factors.

By early classical times, Pythagoras and Plato were preserving the numerical mysteries of the ancient and megalithic worlds for a future where decimal notation, the alphabet, arithmetic, and literacy were revolutionizing the human relationship to numbers and language. Numeracy was still considered to have its roots in the cosmological, because religion had been born from the cosmos through astronomy; that is, astronomy was the formal cause* of numeracy: The planetary cosmos was full of numerical forms and coincidences from which theories arose as narratives and through which the types of harmony operating in the world were first identified. The cosmos was a diversity arising out of a unity, seen so clearly in whole numbers finding their origin in oneness and in their a priori invariances, seen as repeated in the sky as if by design. In this, numbers were clearly a manifestation of Will—but whose?—as well as useful for the secular recording of things and for functional arithmetic so that "things add up." Numbers came to be both of these things, but would the issue of numeracy exist without the consciousness of a cosmos that demands numbers be used to discover the otherwise invisible structure of the geocentric world? The creativity of the cosmos was increasingly replaced as Greek creativity unleashed literacy—that is, a culture of personal writing and personal reading.

PYTHAGORAS

Throughout the Mediterranean cities of the Greeks, Greek literacy was a major cause of their famously original thinking. Literacy was a skill where words became, through writing, a dialogue with oneself and others, revealing intellectual possibilities and new subjects. The myth of Pythagoras sees him gathering material for his eventual number systems, in the ancient mystery centers of 600 BCE—such as Egypt, Babylon, India perhaps, and, of course, from the Greek Mysteries—before returning to the island of Samos in middle age. He established Pythagorean schools in coastal cities in southern Italy and Sicily. His own system, alongside contemporary pre-Socratic philosophizing, influenced Arabic thinkers—such as al-Kindi of the Babylonian Hall of knowledge and ibn-Sina—who were part of an Islamic renaissance four hundred years prior to the one in Europe. In turn, later Christian scholars and churchmen would draw on classical and Arabic works in translation that were the prelude to modern science based on knowledge of the physical world, a field originally called natural philosophy and starting with the pre-Socratic philosophers.

---

*This term is from Aristotle, Plato's pupil, who saw how the possibility for something was made up of four quite different causes for it, a *formal* cause leading to something's actual *material* causation though two intermediate causes of skill and design, called the *efficient* and *final* cause, respectively: making a foursome, or *tetrad,* as a structure of Will.

In the east of the Aegean Sea, Samos had historical sons other than Pythagoras: Herodotus, who wrote early histories of the ancient world; the philosophers Melissus of Samos and Epicurus, who believed only the senses are a reliable source of knowledge about the world; and the astronomer Aristarchus of Samos, the first known individual to propose that the Earth revolves around the Sun. To the north of Crete, Samos was off the western coast of Turkey. Pythagoras's father was a successful sea trader, who dedicated Pythagoras to Apollo and the Delphic oracle, naming him after the Python said to have lived beneath the oracle until Apollo killed it.

Greek religion and its temples were also changing. The temple of Hera at Samos (circa 800 BCE) was innovative for its colonnade and the elongated rectangular form that later Greek temples, called "hundred-footers," would take. In the new patriarchal world, Hera was the wife of Zeus, king of the gods, but she is more likely the consort of Herakles (lit. "glory or fame of Hera"), who, having completed twelve labors, is therefore pointing to the role of Zeus-Jupiter, who sweeps through each of the twelve zodiacal signs, on average, in 361 (19 × 19) days each—a year that could be called here a zodiacal year, uniquely still observed as a calendrical feature within Indian astronomy. Figure 5.22 shows how this year was related, by the Cretan goddess culture, to the Venus synod of 584 days in the Fibonacci ratio 55 to 34, something also found in Cappadocian rock-cut churches (see figure 7.3 on p. 149). This figurine could be the goddess Hera. Analysis of the Heraion temple of Samos reduces to the four-square model of the Sun and Moon, implied in Carnac[2] and rendered in megalithic units (see fig. 6.5).

At her Samos temple, Hera is aligned to the northern-horizon events of minimum lunar standstill, whose light was shuttered into the inside of the temple only at the Moon's rising in the northeast at that standstill. To the east a second chamber points directly to the Moon's setting at its northwest minimum standstill. Built just two or three centuries before Pythagoras, and in Samos, her temple's design embodies both the megalithic understanding of time and the future format of stylobate Greek temples. This junction between the past and future lay at the heart of the creativity of ancient Greece.

The Indo-European Pythagoras

The mysteries of Pythagoras's day were made up of ideas that probably arose in different regions of Eurasia and Africa, with each one perhaps isolated from the others.

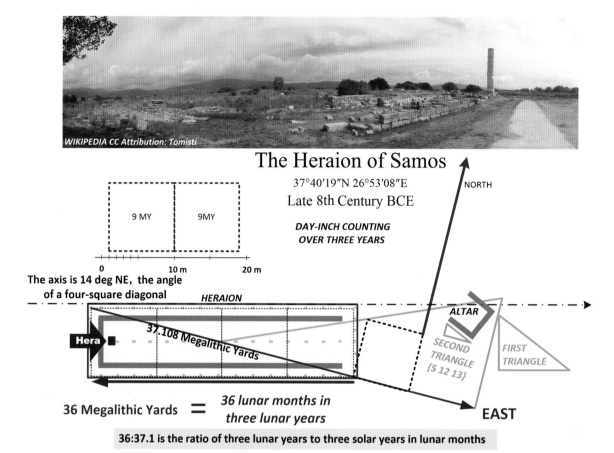

FIGURE 6.5. The Heraion of Samos was megalithic counting platform aligned to the lunar minimum, which she seems to have represented. From Richard Heath, 2021.

1. The tradition of God as Great Mother was evidently expressed by the Minoans and the Maltese megalithic culture. In the sleeping goddess of Malta figurine, correlated with the Olympian Greek myth (see fig. 2.5), the female is receptive yet also primary as the Earth goddess.

2. God as Father is patriarchal, and, owing to the nature of sex, the corollary seen in the Egyptian picture is of a reversal, where the sky goddess was Nut, and the male is looking up to her.

3. God as a sacrificial deity or "Son" is an idea that emerged at the origin of the Indo-European language, where man was created in sacrifice to create, maintain, and save the human world, initially through language acts signifying the higher powers (see chapter 8.)

4. God as Great Spirit appears to have been a development of the East, including Persia, of the direct causation of phenomena by an informing spirit, an intelligence greater and inevitably different from the human. The agrarian religion of the first Zoroaster was of light and dark, and similarly had a great light spirit and his dark nemesis, then rather like the Rg Veda.

J. G. Bennett used these different ideas as an explanatory framework, where these four ideas had developed as a structure of Will, each in a regional isolation until encountering other ideas in prehistory due to climate change. This hypothesis could have possible forensic consequences for the formation of the ancient language groups such as the basic indicative,* trilateral, agglutinative, and Indo-European languages. Using that hypothesis, one can see how the matriarchal and patriarchal collision in the Mediterranean could suit such an explanation. The language of the megalithic, based on numerical length as time numbers, geometry, and aspects of goddess art regarding the creative time-forms, would then be a high-performing *indicative* language of the Stone Age that arose during the megalithic projects at different latitudes, starting with the Arctic, where (if the Vedas began there) a descriptive language evolved (see chapter 8) to describe a polar astronomy quite different from our own. In addition, one can see the subsequent collision of preclassical cultures as the blending of different ideas of God, and their language groups, leading to various outcomes, such as

1. a full adoption of the best descriptive language (quite unlikely);
2. the modification of one's native language, adopting the convenient innovations of the grammar of the Indo-European language (weak);
3. the creation of many dialects, of the best descriptive language, evidenced by the diverse Indo-European languages spoken throughout Europe and Asia and carried through Europe as part of the Neolithic package (as has occurred); and
4. the retention of the indicative language owing to continuing isolation, which prevented blending and protected extant traditions such as foraging and megalithic astronomy.

---

*Bennett called these languages *ostensive,* but the meaning of that has recently changed to "stated or appearing to be true, but not necessarily so," while *indicative* means "serving as a sign or indication of something."

For example, the idea that numbers created the world was first possible when megalithic astronomy used the powers of number to count time and manipulate different lengths of time geometrically, using metrology. The metrology required became a worldwide code for the work of building temples (and for laying out fields, and for other buildings), too. As an exact science of proportion, foot-based metrology evidently reached both patriarchal and matrilineal tribes.

From the above, the language Pythagoras spoke in 600 BCE, as a member of the the northern migration of tribes, was ancient Greek, which had evolved from Proto-Indo-European (PIE), which is a descriptive language. However, many of the mysteries of his time hailed from the matriarchal world, whose language was indicative. Being based on number, Pythagoras was understanding them with a descriptive language, at a time when ancient Greek (1500–300 BCE) was becoming alphabetized, leading to the first truly literate world of classicism. Mysteries played an important role in the classical worlds of Greece and Rome, the Roman world for pagan being rural or rustic. That is, the background beliefs of Europe such as those of the Celts hailed from the Goddess, something that Emperor Julius Caesar found fascinating. The difference between languages is significant because of the way a descriptive language sees the world in a less holistic way, in contrast to the megalithic vision that saw the whole sky as a form rather than as a set of facts. This dichotomy, of types of perception, has been extensively established recently by McGilchrist (2009), whose work *The Master and His Emissary*[3] was summarized on BBC Radio 4 as "about the specialist hemispheric functioning of the brain. The differing world views of the right and left brain (the 'Master' and 'Emissary' in the title, respectively) have, according to the author, shaped Western culture since the time of the ancient Greek philosopher Plato, and the growing conflict between these views has implications for the way the modern world is changing."

We have seen that the ancient gods, derived as they were from a celestial form, could only have been made available through the megalithic, so that megalithism had in part become the Mysteries (appearing in Greece and the ancient Near East) as literacy and reductionism came to define the classical worldview, mixing with the Indo-European carrier wave of the Neolithic package. Pythagoras therefore would have seen the Greeks, and the numerical mysteries, speaking two different languages: (*1*) dialects of Indo-European or other native tongues written down using alphabetic writing; and (*2*) more indicative languages, of the gods and mysteries, glued together by a poetic language of whole numbers, rational fractions, and geometry.

The Musical Harmony of the Cosmos

Pythagoras used the simplest type of dimensional separation to show how musical ratios emerged from the first two primes {2, 3}, as seen in figure 6.6.

Because many gods were planets or astronomical phenomena, Pythagoras declared the planetary mystery "the harmony of the spheres" on the basis, I believe, that musical harmony "heard by the ear" via experiments on the lengths of musical strings using a monochord could explain the tone and semitone relationship between the synods of Jupiter and Saturn and the lunar year as numerical. But the secrecy of his cult appears to have worked flawlessly to lose sight of this meaning, as no trace of it exists—unless it is still secret. More complex relationships have since been proposed as to what this mystery pointed to, all missing the fundamental and countable point that the Moon is the basis of a musical harmony with the outer planets. Geometry helps you see this, but the geometry of the solar system as such only bears deniable evidence of a creator of it such as Plato's Demiurge, which was harmonic in dividing according to number.

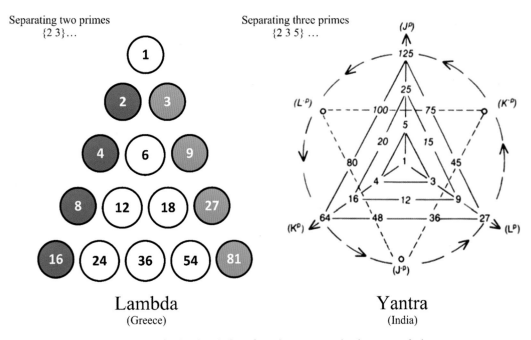

Lambda
(Greece)

Yantra
(India)

FIGURE 6.6. *Left,* the lambda of Pythagoras, which ignored the prime number 5 as a dimension of harmonic development; *right,* triangular yantras grasping all three early primes, each one like arithmetic "holy mountains" focused on actual octave ranges (see fig. 6.3) to map the factors {2, 3, 5} in two dimensions.

However, it cannot be proved Pythagoras and Plato (see p. 128) inherited knowledge of the musical intervals between the outer planets and the lunar year or, if so, from whom. Pythagoras has therefore been declared the first in recognizing the primacy of the octave's division of the oneness of the ecliptic. For us, Pythagoras is the father of the single scale formed by {2, 3}, which he preferred as being divine, to the seven possible scales deriving from {2, 3, 5}, perhaps because popular music is naturally tuned to this "by ear." Modern music has now replaced scales with keys, tonality, and the theory of harmony only possible with equal temperament, involving the division of the octave (as Zeus might) by twelve equal semitones, like the twelve mean solar months of the solar year.

However, in planetary harmony the two intervals between the lunar year and two outer planets, as 9/8 and 16/15 requires {2, 3, 5} when seen as an octave between 9 and 18 lunar months, as employed by the Olmec tradition in Mexico. This required just intonation for a planetary tuning theory, as it was in Mesopotamia, which predated Pythagoras by two millennia. There is the distinct possibility that the octave was a deeply religious secret or mystery, belonging to the priestly classes of the ancient Near East, for its religious interpretation of the cosmic reality based on {1, 2, 3, 4, 5, 6, 8, 9, 15, 16, etc.}, their five ratios {2/1, 3/2, 4/3, 5/4, 6/5}, and the tones and semitones {9/8, 10/9, 16/15}.[4]

THE PARTHENON

The Parthenon was built two centuries after Pythagoras, and, resembling the Heraion of Samos, it embodied the cosmic knowledge of the Moon, Jupiter, and Saturn in a temple devoted to Athena. She was a matriarchal goddess cut from (or recycled through) Zeus's patriarchal head, after he had swallowed the goddess Metis (lit. "wisdom," "skill," or "craft"), who was his first wife (see fig. 4.3). In Greek myth, she is laundered as "an Oceanid, one of the nymphs who were the daughters of Oceanus and his sister Tethys and who were three thousand in number. She was a sister of the Potamoi (river gods), sons of Oceanus and Tethys, who also numbered three thousand." A perfect description of a large matriarchal tribe.

Metis was the first great spouse of Zeus, himself once titled Metieta (Ancient Greek: Μητίετα, lit. "the wise counselor") in the Homeric poems. Metis was a boon, being the one who gave Zeus a potion to cause Cronus to vomit out Zeus' siblings, but she was a threat because she would give birth to a son more powerful than Zeus, who would eventually overthrow Zeus and become king of the cosmos in his place. Zeus tricked her into turning herself into a fly and promptly swallowed her.[5]

Probably, therefore, it was the goddess culture that had given Pythagoras knowledge of the outer planets' musicality with the Moon, and later Pythagoreans, including Socrates's father, created the Parthenon to express and preserve this secret knowledge. The Parthenon is Athena's revenge for eating that fly. The numbers used in the Parthenon belong to an octave from 720 units to Adam's 1440 units (see fig. 6.7).

Prehistoric number science was holistic and only practiced as a form of *understanding,* rather than *knowing.* Nothing of material benefit necessarily accrues from understanding, whereas civilization has used knowledge to achieve convenience and luxury. Correspondingly, it is remarkable how little use

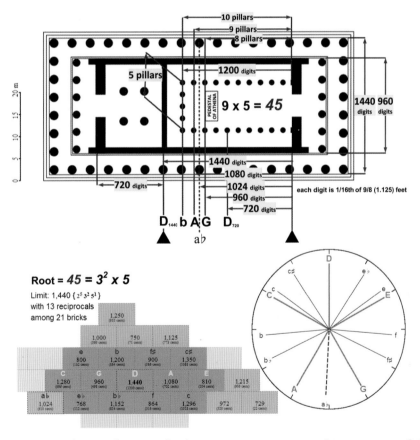

FIGURE 6.7. The Parthenon of Athens as a cosmic musical instrument of the Sun, Moon, Jupiter, and Saturn.

Pythagoreanism has been: endless generations of intelligent, usually privileged young men became expert in it, but, in the end, modern science has found little in it except as an antique basement for pure mathematics because, as already stated, Pythagoreanism lost the factual basis of its "harmony of the spheres."

PLATO'S *DIALOGUES*

In the new world of Greek literacy, philosophy of many sorts became possible, outside the traditional power of professional reciters of oral epics and myths. Pre-Socratic philosophers operated like letter writers, sending copies of manuscripts for others to read, often in other Greek cities around the Mediterranean. Socrates was a staunch believer that the oral tradition had throttled the human ability to use language to think for oneself, and he pointed out that language use was in this way unconscious and was being used in a regressive fashion to merely remember and repeat the same old stories. Socrates's father was a stonemason who had worked on the Parthenon when he was growing up.* His outspoken opinions about truth, veracity, and logic led to the Socratic method of teaching through successive questions and answers until the respondent showed that they were already able to state the truth about something but had not asked the right questions of themselves. The authorities finally saw him as a threat and poisoned him, in part because he was disrespectful to the old gods of the oral tradition.

Socrates himself became a major character in a set of Plato's dialogues between various characters; the dialogues were intellectual dramas about many subjects dear to this new offshoot of Pythagoreanism, where individuals were now free to write and think outside of the oral tradition. But at the same time, writing also saved many archaic oral works that were written down for posterity without needing oral performers to recite them from memory: one could read them oneself, silently, or aloud to others. This was an emancipation of information and an individualization of thinking. Like a tar ball, it drew a range of subjects into its orbit. These included elements of the megalithic number sciences, where numbers could be letters, and thoughts about musical harmony and its role in the creation of the geocentric world could be defined by a

---

*Note the similarity to the prophet working on the repair of the Kaaba, whose form is quite similar to the single-aisled churches of Cappadocia. Not a writer, we rely on Plato to learn of Socrates. The *Dialogues* often feature Socrates and his teaching-through-questions and quite modern attitudes of learning from what you already know.

new character: the Demiurge as a creator working with numbers to define the Earth and planets in space and time using musical intervals, as Plato discussed in his *Timaeus*.[6] The governance of city-states could be compared to different tuning systems, including the fabled Atlantis—a subject very relevant here.

Plato's singular story of Atlantis may be pointing us to the megalithic and Malta's ideal situation overlooking the sea at 36 degrees north, the extrapolated location of the earliest constellation mappings around 2800 BCE, and the subsequent demise of the palace periods of Crete and the matriarchal Bronze Age world, soon after the volcanic caldera of Santorini exploded just north of Crete.

Like Malta and Crete, Atlantis was an island with mountains to the north (mainland Europe) and a large rectangular plain* (the sea perhaps), and Poseidon's artificial island (perhaps the southern one-time polestar Canopus) inland of the island's southern coast (see fig. 6.8). The whole arrangement was described quite geometrically, and so it was probably coded as a vehicle for the presentation of some ancient knowledge. The word used for *plain* by the Pelasgian Greeks largely meant "flat sea," because Greece had lost most coastal plains owing to the inundation of its coastlines after the Ice Age. Harald Reiche

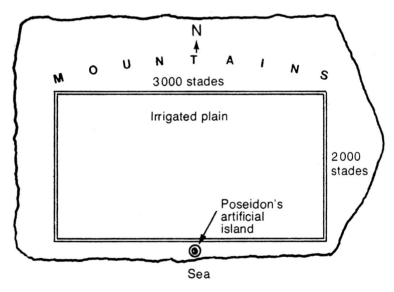

FIGURE 6.8. Plato's island of Atlantis, given to Poseidon when the gods apportioned heaven and Earth among themselves in the Golden Age. After Reiche, "The Language of Archaic Astronomy."

---

*Measuring 3000 stades east–west and 2000 stades north–south

interpreted the island as having been an astronomical tableau with harmonic aspects, and Poseidon's Island[7] is emblematic of Saturn's golden island (where Zeus exiled the Golden Age ruler Saturn) and the literally golden southern polestar of Canopus (El Ponderosa, or "heavy one") often confused with Saturn because Saturn is slow in its thirty-year orbit. Canopus only just rose for the Minoans at 35 degrees north, just visible to the south and above the plain of the Mediterranean Sea, a plain that then represents the sky view *above* the sea horizon, as the sea of stars.

Atlantis might therefore have been a fabulous reference to Malta's astronomical view over the Mediterranean and the destruction of the matriarchal culture of the Minoans, destroyed by the massive eruption of Santorini sometime between 1645 BCE and 1600 BCE and matriarchy's continuing displacement by patriarchal, northern tribes.*

The explosion of Santorini in the mid-second millennium BCE, on the island now called Thera, was a factor in the end of the Old Palace period. A city on Thera called Akrotiri survived but was empty of human remains under a blanket of ash, and the place seems to have been a cult center for the Minoans. It is 73 miles north of Malia and 73 miles north again is Delos, the birthplace of Apollo, the sun god and subject of a continuing patriarchal cult whose chief shrine was the oracle of Delphi, which, with Athens in between, all lie on the midsummer sunrise to midwinter sunset line at the notional 30 degrees SSE to NNW (see fig. 6.9).

Relatively little dust landed on Crete, but a large tsunami hit the northern coast with great damage and loss of life, and a significant number of Minoan ships probably sank or were wrecked. There were also earthquakes, and it is thought that soon after, Crete came under the influence of the Mycenaeans, a patriarchal tribe from the Peloponnese who invaded Crete with boats and created the Late Minoan period, the new palaces having again been destroyed.

By 1200 BCE, Iron Age adventurism appears to have destroyed most of the eastern Mediterranean cities and powers, and Greece fell into a "dark age" only illuminated for us by texts written down from oral sources as the new alphabet was applied to recording oral works such as Hesiod's cosmogony, Homer's epics, and other works that together contained a coherent set of divinities and myths we call the Greek myths, which is remarkable on its own unless these oral works

---

*Many have thought this to be the simplest explanation of Atlantis, but we will see in chapter 8 that the Arctic is also a contender.

FIGURE 6.9. The alignment of Delos, Akrotiri, and Malia.

came from another coherent cultural source, partly Minoan, perhaps Hittite. But our search for the form of history cannot linger here.

THE BIBLE AS A NEOLITHIC TALE

The Bible's first books, written around 600 BCE during the Babylonian Exile of the Jews, borrowed and adapted well-known Mesopotamian myths to create various "Neolithic" stories, among them:

Adam came after the creation of the Earth and sky.

Eve, the first woman, lived with Adam in the earthly paradise of Eden, from which they were expelled.

They had two unlikely children, Cain (a farmer), who kills Abel (a shepherd);

Noah built an ark to survive a divinely authorized flood that would otherwise have wiped out humanity.

God destroyed the tower of Babel, which would reach unto heaven, causing people not to understand one another.

Those people plus several long lists of progeny with rather long lifetimes, get us to

Abram, who was a patriarchal person from Ur of the Chaldees (near Sumer), a very Neolithic context and place of origin.

The Lord God renames Abram Abraham, who is to found the patriarchs through his childless wife, Sarai, renamed Sarah.

Sarah conceives Isaac at 90 years old and he then lives to 180 years; that is, to double her age.

In the beginning, there was darkness, and God said, "Let there be light," a widespread cosmogonic theme because light comes from the Sun and sky by day and from the stars and planets at night.

The six active days of the creation resemble the first six whole numbers {1, 2, 3, 4, 5, 6}, which give rise to the key intervals of musical harmony—namely, the octave, fifth, fourth, and major and minor thirds.

The number 7 does not enter into harmony and so, it seems, the seventh day was a "day of rest," for appreciating the creation as "good." But the music being created is not for the ear but rather to establish harmony as a feature of cosmic design and its primacy. The early chapters would later be followed by Abraham and his children, who rapidly become the twelve children of Abraham's grandson Jacob, renamed Israel. The musical harmony of seven note classes in an octave also has twelve unequal semitones* who became the twelve male Children of Israel.

The number of Adam, as 45, is the harmonic root of six octave doublings {45, 90, 180, 360, 720, 1440} (imaging {1, 2, 3, 4, 5, 6}) with a seventh doubling giving 2880, the onset of eclipses and twin tritones within the patriarchal octave of the twelve tribes. The number 720 or 1440 in limit enables only five modal scales, while 2880 enables seven scales having full chromaticism with tritones.

In contrast, Plato's tyrant (729)—tuning by fifths—produces, after twelve such fifths (numerically the limit $16 \times 729 = 746{,}496$), a span of seven octaves but then producing two audibly disharmonious tritones.

The key to Adam's fecundity was the number 5 (equivalent to the Hebrew letter *hey*) as in $45 = 9 \times 5$. Using prime number 5 improved the harmony with a small number (2880) as the limit.

The new theme of octaval number theory was obviously used to prefix a history of the Hebrews. This first-millennium-BCE innovation in numerical tuning, allied to the use of alphabetical written letters of the Hebrew language, allowed each letter to stand for a number in decimal as in {1, 2, 3, 4, 5, 6, 7, 8, 9, 10, 20, 30, 40, 50, 60, 70, 80, 90, 100, 200, 300, 400}.

Humans have a deep relationship to the number 5 as four limbs plus head and four fingers plus a thumb, used for counting in the decimal fashion, which the Jews adopted rather than base 60 since (*a*) it was simpler; but also, (*b*) the letters of their new alphabet could have number-letter association as with Adam

---

*Modernity has made semitones into equal temperament where equal semitone intervals of the twelfth part of $\sqrt{2}$, multiplied by 12, gives an octave, no interval then being rational, a compromise to allow changes of key, musical ensembles, and classical music.

or ADM, which equals 1 × 4 × 40 = 45; and (*c*) the Lord God did not like what the Babylonians did with large numbers, which as gods gave him a headache. For example, they made bricks with water and mud and built a ziggurat that reached heaven, repeating the motif of Adam being thrown out of Eden for eating the fruit of the Tree of the Knowledge of Good and Evil. This idea of repeated imagery in myths adapted from Babylonia and acted out by the patriarchal players could provide a context for where the Hebrews had come from, and the small decimal numbers could give access to a better tuning theory, suitable for humanistic music. It seems likely that Eve gave Adam the fruit from the tree because she stood for the Mesolithic goddess culture that had performed, according to our thesis, megalithic astronomy. That astronomy touched upon heaven like the ziggurat, which was conceived to be a model of the Earth itself, its seven latitudes. The Great Pyramid was of a height commensurate with the polar radius of the Earth, the pole and celestial heavens being a common location of God's heaven in many traditions.

The second major theme, obviously numerical and harmonic, was that Adam's number 45, multiplied by 16, is 720, which is the earliest number capable of supporting the five scales according to the Jewish decimal version of Mesopotamian tuning methods. The octave, also important to Pythagoras (who was contemporaneous with the writing of the Bible), was important to the astronomy of the Mesolithic astronomers because they had seen the simple musical tone (9/8) and semitone (16/15), between the Jupiter and Saturn synods relative to the lunar year of twelve lunar months, Saturn's harmonic root being 1, or unity; that is, Saturn was the cornerstone of a harmonic planetary system. The fruits of the north polar tree hung above the Sun's path and were part of a large set of numerical coincidences modeled on the structure of the number field, as seen in the case of how the first six numbers define musical harmony and how seven units as a diameter creates twenty-two units of the circumference of a circle, also the number of letters in the Hebrew alphabet.

Furthermore, the singularity of the one god (though tribal)—as opposed to the many gods of the Mesopotamians and Egyptians—had the direct consequence that musical harmony belonged to one divided into two parts, two into three, three into four, and four into five, and five into six; thence the creation was using numbers that only seem abstract until they manifest as lengths of time that then interact with one another, as time intervals between synods. And, at this point, the astronomy could be packaged within the lunar calendar of twelve lunar months, known especially as a Semitic calendar.

Another idea apparently borrowed from the Mediterranean goddess was the Saturnian calendar based on seven days in the week. The Jews suppressed the obvious precursor to this week in which the planetary names were given to Monday (Moon), Tuesday (Mars), Wednesday (Mercury), Thursday (Jupiter), Friday (Venus), Saturday (Saturn), and Sunday (Sun)—through an Egyptian system based on daily planetary influences. All of the Indo-Europeans have the planetary version, probably because the seven-day week had traveled as part of the Neolithic package and the evolving regional forms of Proto-Indo-European, causing a plethora of dialects and adaptations now called Indo-European. The day of rest was the Sabbath, traditionally the day of Saturn, whose synod is 54 weeks long and whose year of 364 is 52 weeks long, only now the days were called 1, 2, 3, 4, 5, 6, and the Sabbath was 7, coequal to Saturn's day but as a number. The original Jewish menorah had seven candlesticks, including the middle trunk, a tree form.

One of the most significant introductions of the Neolithic in Genesis lies at the end of the seven-day creation. Humans were made "in the image and likeness of God," which sublimates the Indo-European belief in a god made flesh as a necessary sacrifice to rid the human world of its sins. In the Puranas it was Martanda (sometimes an extra eighth *aditi,* or "light-bearer"), and the Sumerians also sacrificed a lesser god Kingu,* who, like humanity (as Adam), had been given the Tablet of Destinies by his mother, Tiamat. It is this semidivine nature of humans that led to the desire for a savior or messiah and the life story of Jesus, as son of God, in the New Testament. But this idea was also then present almost universally; that is, in the abstract as derived from oneness. It points to the human having been created for a definite reason, sometimes explainable harmonically as the need to compensate for the inherent tuning errors in the planetary tuning system, and the harmony of the spheres that obviously shares these errors. Orpheus was another such figure.

All this perhaps explains a Jewish Jesus born to a Virgin Mary, a transposed

---

*From Wikipedia, "Kingu, also spelled Qingu, meaning "unskilled laborer," was a god in Babylonian mythology and the son of the gods Abzu and Tiamat. After the murder of his father, Abzu, Kingu served as the consort of his mother, Tiamat, who wanted to establish him as ruler and leader of all gods, before she was killed by Marduk. Tiamat gave Kingu the Tablet of Destinies, which he wore as a breastplate and which gave him great power. She placed him as the general of her army. However, like Tiamat, Kingu was eventually killed by Marduk. Marduk mixed Kingu's blood with earth and used the clay to mold the first human beings, while Tiamat's body created the Earth and the skies. Kingu then went to live in the underworld kingdom of Ereshkigal, along with the other deities who had sided with Tiamat."

Mesolithic goddess figure, in a matrilineal culture where men do not have patrilineage but carry their mother's name instead. Also, the virgin is the triple goddess Venus, and Virgo was the constellation ruling the pillars of the year in the Virgoan precessional "month," which was during the Ice Age. The antetype of Jesus is seen by many in the near sacrifice of first-son Isaac by Abraham as well as in Joseph (one of the twelve male children), who fulfills the story of Abraham's original journey to Egypt, under famine, where later the children of Israel, swollen to 600,000 persons, had become enslaved so that Moses, by stealing a prince's education, led an exodus to the Promised Land, YHWH being $6 \times 5 \times 5 \times 5$ is code for $6^5 5^5$ (using powers), equaling 777,600,000—that is, Moses (777) leading the Israelites.[8] Eden becomes transformed into the Promised Land, given to the covenanted people called Israel, God's new name for Jacob. His role complete, Moses could look upon the Promised Land but dies before entering it, though he is the first to know God as YHWH.

Before the Bible, no one had attempted a large history, set within myths and legends, that came to identify and still mobilizes a whole people. The new Mediterranean tool of alphabetic writing, decimal arithmetic, literacy, and intellectual mastery of Egyptian and Babylonian number sciences, could incorporate the narrative techniques of the oral storytelling and harmonic allusion to create an integrated written document with many levels of meaning enshrined within religious buildings as scrolls, with a calendar of festivals and norms that differentiated the Jews, even after their diaspora outside the Promised Land.

The two key sources for Christianity and Islam, the later peoples "of the Book," became the Bible and the Greek works being written during the same epoch. From these, the Western tradition was formed, and so it is no wonder that Saint John started his Gospel with the words "and in the Beginning was the Word," because Light would ambiguously allow the Mesolithic world to be seen and quantified, but the Word enabled the world to be understood as an object for creating meaning, created by God. God was first called El Shaddai, translated as "Lord God," even by Abraham, until the name YHWH, often translated as "Jehovah." Ernest McClain pointed out that just as Adam can be seen through gematria as 45, YHWH as 10.5.6.5 is 26, yet Adam could be seen as decimal place notation 1.4.40 or 1440, 32 times his root of 45, and YHWH can be seen as $60^5 = 777,600,000$, a very large harmonic number that stops just short of the even larger number of Indra and Marduk of 17,280,000,000 which was predicated on the $\sqrt{2}$ and which set the cosmic machinery in motion (see next chapter about Indra's creation of the Earth's obliquity). YHWH chose the middle path, of powers of 60, to define how the human essence class could be in his own image and likeness. His was the path of Anu to the power of 5 or the letter *hey.*

A fuller treatment of this subject, with a full introduction to the numerical musicology discovered by Ernest McClain, can be found in chapter 3 of my *Harmonic Origins of the World.*

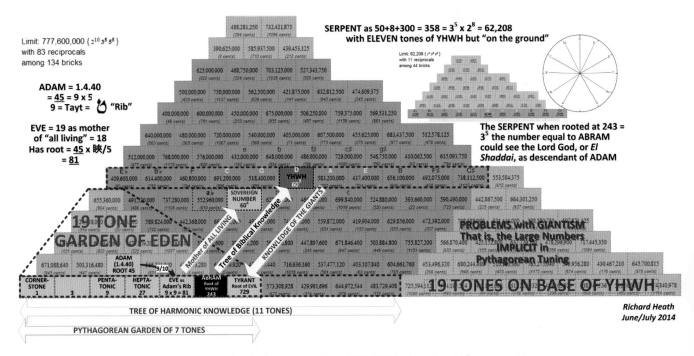

FIGURE 6.10. The holy mountain of YHWH showing Eden as a 19-tone limited by 1080, just as the mountain has 19 tone numbers along its base.

PART THREE

THE ROLE
OF
PROVIDENTIAL
HISTORY

By the end of the first millennium CE, dogmatism in both Islam and the Christian Church would keep new thinking in check for centuries. But centers of learning were still focused on recovering whatever was left from classical times of the many original works lost, some then censored.* Classical Greek contributions to scholastic orthodoxy and the popularity of Greek and other myths with men of learning meant there was still a great literature of remaining texts within which literary reference was allowed for the purpose of systematizing existing ideas, without inventing wholly new ones or looking for anything new. Hence, one could not really do astronomy except through scholarship involving past astronomical literature and tables of data. Cuneiform was not yet readable, and, in any case, the clay tablets containing it awaited the birth of archaeology, philology, and so on.

The planetary model of the cosmos had long been evolving until, in the medieval world and the second millennium, the simplified "Chaldean" geocentric model was the cosmos. The spherical version had Earth at the center of planetary orbits that were seen as attached to spheres supported by a hypothetical and transparent substance called the aether and beyond which lay the prime mover of them. The planets were visualized as placed upon seven equally spaced shells, and each carried one of the sometimes-wandering planets visible to the naked eye. The nearest planet was the Moon, as it revolved around Earth, then the inner planets Mercury and Venus, which moved around beside the Sun, then the outer planets Mars, Jupiter, and Saturn, and, finally, the sphere of the fixed stars, including the zodiac, whose motion was in some way driving all the planets from a primum mobile. The physics we experience on the surface of the Earth and the atmosphere above it were accounted for by observing the four states of matter as four elements constituting the Earth—namely, solid things including earth, liquid water, gaseous air, and plasmic fire. These elements constituted a sublunar world, in the space between the Earth and the

*For a brief period at the end of the first millennium, Islam's natural philosophers broke free, but this was brought to heel through persecution, imprisonments, and even executions.

Moon, while all else was considered "above" that, or metaphysics. By the current era, this model was conceived of as the spiritual path back to God, who was beyond the primum mobile, since by the period of new teachings and teachers around 600 BCE, individual human development toward a single creator became the emerging message, governing—in Bennett's terms—an approaching Megalanthropic Epoch that, in 600 BCE, was displacing the Hemitheandic Epoch and its man-god kings.*

But during the early centuries of the current Christian era, and long before the entrenched centuries of the medieval church rolled on, Christianity was scattered into sects whose beliefs did not correspond to any one common doctrine on matters of great substance, such as whether Jesus was the Son of God, or whether God had entered Jesus at his baptism, or was he a new type of enlightened man who would evoke the transformation of others into what he had become? The problem was exacerbated in that many would-be Christians came from groups holding Gnostic opinions on such ideas. A showdown occurred in Cappadocia, between Rome and Jerusalem, where a group of three "fathers of the church" fought for the idea that Jesus was God incarnate. This required a doctrine that God contained three aspects in a trinity of Father, Son, and Holy Ghost, since Jesus, the son of God, had said a comforter would come, the Holy Ghost, though the one God lay behind all three of them, and the duality of man and God was reconciled through a third aspect, the Comforter. The resulting Nicene Creed was then adopted by the new Eastern (now Orthodox) Church, coincidentally created by the Roman emperor Constantine in 300 CE. When the Church of Rome subsequently adopted this creed, Christians everywhere had to accept it to belong to the true faith and its Pauline Church of Saint Peter in Rome. This established the "fundamental image myth" of the Roman Church.

But in the same region of Cappadocia, an alternative activity was soon being carried out to save the ancient astronomy of the Mesolithic past, perhaps under the rubric that Jesus was now Lord of the World. By 500 to 600 CE, the volcanic bedrock of much of the region was being used to create monasteries safe from border skirmishes with the Persians and then with Islam; the latter would eventually prevail in the form of the Turks converted to Islam. Whole

---

*Megalanthropic meant that now man would be great as is seen in the litany of men's names who discovered things. Bennett's work on "History" in volume 4 of his *Dramatic Universe,* supplemented the doctrine Precessional "Ages" of 2160 years (such as "of Aries" and then "of Pisces") with epochs of variable size to express these different dominating ideas.

villages were created underground, reminiscent of the Roman catacombs but also of Crete and quite similarly of Malta and Sardinia. In fact, the original Cretans are said to have come from Anatolia. Truly an underground movement, churches were built using inner dimensions that stored, recorded, and seemingly extended the megalithic knowledge of astronomy. Outwardly orthodox, these churches were hermitages, perhaps to a nearby monastic order, often occupied by an advanced hermit monk with only occasional use for group meetings.

While outer Christianity was consolidating its doctrinal hold over what it was, an alternative story belongs to Cappadocia as a literally underground phenomenon, existing to this day while normal buildings have disappeared.

7

CAPPADOCIAN CROSSROADS

IN ANATOLIA

The current era is defined by the advent of Jesus, yet his existence was unknown to all but those who had met him, or those who had contact with them during centuries of persecution by the Roman Empire of their sect, which was established by a vision of Jesus by Saint Paul, who came from near Cappadocia, to be the chosen bringer of Christianity to the Gentiles. His historic influence upon the future cultural life of Europe came through the formation of the Roman Church of St Peter, his writings to the early churches and arising of the New Testament with its four Gospels, his letters and works of other saints constituting the Christian Bible. Before that, during the first centuries lists appeared of the sayings of Jesus, perhaps heretical; these were either inserted into Gospels or lost until recent times. The Gospels, the stories of the life and mission of Jesus, consolidated both his historical existence and his teachings. The time between his birth and his death and resurrection was traditionally and astronomically significant, at 33 years, which is the solar hero period of 12,053 days, but each of the Gospel biographies differs in its approach and content.

EARLY CHRISTIANITY AND THE FOUR GOSPELS

The gospels of Mark and Matthew were lections, organized around the solar year as a set of forty-two liturgical readings, in the case of Mark, for each Sunday in the year with the readings for the remaining ten other weeks of the year coming from other texts in the Old Testament or other main books of the canonical New Testament.* The Gospel of Mark was probably already translated into

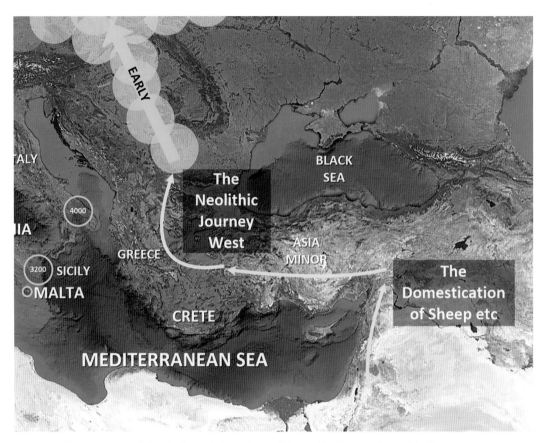

FIGURE 7.1. The influential region of Anatolia in southern Turkey, once called Asia Minor, lay on what had been the Neolithic path of diffusion to the east coast of Greece, which was now between Jerusalem and the Holy Land, the emerging Eastern Church, and Rome.

ancient Greek by Saint Mark when he arrived at the Greek city of Alexandria, from the east. The Gospel of Mark proved popular in Rome, in part because it presented Jesus's mission as if it started with his baptism and ended with his death on the Cross. By implication this suited the *Docetism* doctrine, that the spirit of God had only entered Jesus at his baptism, when the skies opened up and God's spirit entered him. He then withdrew to the desert for forty days.[†]

By leaving out the Nativity and the Resurrection, Mark placed both Jesus's

---

[*]See Philip Carrington, *The Primitive Christian Calendar.* I draw on this to give a picture of the early Christianity leading up to the Cappadocian rock-cut churches.
[†]A time symmetrical, therefore, with the forty days of Lent before Easter, marking the crucifixion of Jesus.

baptism and death at Easter, as if his ministry (like Mark's lectionary) lasted a single solar year. This view, given Jesus's age of thirty at the start of his ministry, contrasts with the traditional view of Jesus being thirty-three at his death after a three-year ministry. This seems a key difference between the Gospels of Matthew and that of Mark, though pragmatically the extra sections in Matthew and not in Mark were then available as lectern readings for Christmas, his nativity, and Easter, his death.*

The Gospel of Matthew appeared in the Greek language at Antioch, a major coastal town in eastern Anatolia, on the Aegean Sea. Anatolia was a land bridge from the Holy Land to Greece and thence onward to Italy and Rome or, for the Neolithic millennia before, north to the Hungarian plain. Saint Paul's first "book of letters," the Acts of the Apostles, was accompanied by the Gospel of Luke, both of which seem largely borrowed or influenced by Mark, and with fragments from one or two unknown Gospels. Seven of the churches addressed by Saint Paul in Acts were in Anatolia, and Paul himself was from the city of Tarsus, in the southeastern region called Cilicia. He was formerly called Saul, a Pharisee who was also a Roman citizen. Saul of Tarsus, as he was then known, was gratuitously hunting down Christians until he saw a great light that addressed him in the person of Jesus. He was blinded by the experience until, three days later, Ananias of Damascus was directed in a dream to restore his sight, and he then renamed him Paul and gave him the mission to support the coming formation of the non-Jewish members of the Christian Church, who were termed gentiles.

This now-first Gospel of Matthew was probably written in Hebrew, by Saint Matthew himself, and the Hebrew Matthew might have been the source of the additional content then found in the Gospel of Matthew, when content similar to Mark and Luke is removed from it. Our Gospel of Matthew is therefore like Mark in its being designed as a lection for Sunday readings but with Advent and the Passion now included.

The practical need for texts having a liturgical and temporal function is close to the many structural mechanisms found within other texts based on ancient works that use cyclic or other ascending or descending (pedestal) structures, and similar phrasing or repeated themes, to link those sections

---

*Before 150 CE, Easter must have been a fixed date in the calendar, as it still is in the Orthodox Church. It was only in 664 and the synod of Whitby, in northeastern England, that Easter became a "moveable feast" difficult for non-specialists to calculate as "the first Sunday after the first full moon after the Spring equinox."

as corresponding but not necessarily successive in time. Traditional texts, including the Hebrew Bible, also have such structures within their texts, especially in the wholly different fourth gospel of Saint John, for whom in the beginning was the Logos, not the light of Genesis. This difference may correspond to new thoughts belonging to Greek literacy as a creation now made up of words and sacred numbers. Our expression "dark age" only means an age short of written records (and people who can write and read), hence the need for early Christian monasteries during the dark ages after the Roman Empire.

The Gospel according to John appears Platonic in its detailed use of numbers, such as the 153 fish caught by the disciples by casting their net on the right side, a number that points to the four-square rectangle's diagonal when the sides are 12 and 3, since 153 is the sum of the squares 144 and 9, a clue for what is to follow: the geometrical and cosmological traditions of the megalithic and the lunation right triangle with sides $\{3, 12, \sqrt{153}\}$, with $\sqrt{153}$ equaling the 12.368 lunar months in a solar year; a net being a word for a square root; and Jesus evidently reconciling the lunar year of 12 lunar months and the solar year of 12.368 lunar months.[1]

In fact, the Eastern Church was Johannine rather than Pauline. That means it was aligned to John the apostle (later conflated with John the Elder and the John the Baptist) rather than aligned to Saint Paul and Rome. The Roman Church was most happy with the chief syntonic Gospel of Mark, whose structure is a sophisticated ring composition. The split between the Pauline Church, centered in Rome, and the Johannite Church, centered in Constantinople, was perhaps the last natural doctrinal schism the Creed did not detect, once the singularity of Jesus's life was embraced by the Roman emperor Constantine (272–337) and his wife. They created the new Eastern Church in what was a survival strategy, as the Roman Empire was declining and within a century collapsed. But the principality of spiritual power called the state of Rome and its Roman Church would slowly grow in power.

Between 120 and 150, the Gospels of Matthew, Mark, Luke, and John became the only official biographies of Jesus. But the churches would never shake off their more esoteric roots. In his third book against the heresies, Irenaeus gives the appearance of John the Baptist (in the beginning of Mark) as an image of a Lion or Leo and made the order of the Gospels of Matthew, Mark, Luke, and John what it is today. Seemingly out of nowhere, he uses the ancient animal characters of the four-square directions or pillars of the year, characterized by the zodiacal names of the constellations within which the Sun sits at winter solstice, spring equinox, summer solstice, and autumnal equinox. These are the "living things" found in Hebrew liturgical symbolism called the *hayyoth* central to Hebrew synagogue floors, modeled upon Ezekiel's four living creatures.

> The word of the Lord came to the priest Ezekiel son of Buzi, in the land of the Chaldeans by the river Chebar; and the hand of the Lord was on him there. As I looked, a stormy wind came out of the north: a great cloud with brightness around it and fire flashing forth continually, and in the middle of the fire, something like gleaming amber.
>
> In the middle of it was something like four living creatures. This was their appearance: they were of human form.
>
> Each had four faces, and each of them had four wings. . . .
>
> As for the appearance of their faces: the four had the face of a human being, the face of a lion on the right side, the face of an ox on the left side, and the face of an eagle. (Ezekiel 1:3–6, 10)

These are also found in reverse order in Revelation.

> And before the throne there was a sea of glass like unto crystal: and in the midst of the throne, and round about the throne, were four beasts full of eyes before and behind.
>
> And the first beast was like a lion, and the second beast like a calf, and the third beast had a face as a man, and the fourth beast was like a flying eagle. (Revelation 4:6–7)

The problem, however, is that these constellations belong to the four gates of the year around 4000 BCE, and when the spring next sits in one of those pillars, it will be at the spring equinox in the Age of Aquarius. If the assignations are correct, then Eziekiel's vision was (by definition) either two precessional ages, around 4320 years in the past from Jesus's birth or in the Aquarian Age, 2160 years into the present era; that is, almost now. It seems odd for Irenaeus to be letting in this interpretation, except that these four faces of the Gospels might have appeared to him since *they were already in use on the front-page tags of the very Gospels themselves,* used for identifying each book.

THE CAPPADOCIAN CREED

With the adoption of the fourfold canon of Gospels, we end up with an implicitly Docetist document in the Gospel of Mark and the Gospel of John, appearing like a work on Greek number science worthy of Pythagoras or Plato. The leaders of the church had a central need to form an agreed description of what constituted Christianity as a unified religion. The Nicene Creed was a work leading to Christian theology, and the circumstances leading to that came out of the Eastern churches of Anatolia.

G. I. Gurdjieff, according to J. G. Bennett, said Cappadocia displayed profound cosmic secrets within its church "liturgy," by which he appears to have been referring to Cappadocian church architecture, its measures, geometry, and ritual arrangements (see next section), and this led me to wonder what that was.

Cappadocia became the home of many early Christian churches and monasteries without masonry; rather, they were cut out of the soft volcanic rock of that part of Anatolia, to create sacred spaces. Stone-built churches of that time and region have since been robbed for their stone, adapted for other uses, improved, or obliterated, but rock-cut buildings were hidden within the natural landscape and far from the beaten track, and protected when Cappadocia became a militarized border zone, first with Persia and then with Islam. Entire monasteries and towns were built underground, and churches were cut into volcanic pinnacles—cones formed by erosion—and with only incidental patches of arable land but, surprisingly, accessible water courses beneath the ground or in hidden chasms.

FIGURE 7.2. The earliest signs of monastic activity in Cappadocia trace to the fourth century, when small anchorite communities started to inhabit cells hewn in the rock, often eroded into pinnacles, as seen in at Göreme.

These rock-cut churches became famous because of their rare Christian art. The rock protected unique examples of tenth- to twelfth-century Christian art that had been lost in the aboveground stone-built churches. But before that art, the Orthodox world suffered a period called the Iconoclasm, in which religious imagery was banned as idolatrous, and so destroyed.* Its painters were tortured, even mutilated, to prevent their work. Strangely, this was after representational art was first banned in the recently created Islamic world, now the chief threat to the Byzantine Empire when Cappadocia became a militarized border province.

In the twentieth century, interest in these churches was largely for their surviving art and for the internal architectural layouts of the churches when first hollowed out. But interest in their design grew, because the early liturgical features preserved details of how the rites of the early Orthodox Church had evolved. Their liturgy in the vital centuries after the creed was formalized, especially details concerning the Sacrament. However, it is more difficult to date the churches than their art.

The single-aisle churches are thought to be the first rock-cut designs, many built between the "Early Christian" period (defined below) and the Iconoclasm. The evolution of their design was a single aisle of a naos (nave) and threshold to an apse, with troglodyte and alcove furnishings within these two rooms. The evolution of liturgical functions is the only clue to the age when they were constructed. Late twentieth-century interest in their rock-cut architecture led to scale plans being made so that one can look for the use of metrology within them and perhaps even find late links to the 12/77-foot subunit in use at Göbekli Tepe, Malta, and Crete.

SECRETS WITHIN ORTHODOX CHURCHES

The elimination of a plethora of early doctrines (Docetism, Montanism, Adoptionism, Sabellianism, Arianism, Pelagianism, and Gnosticism) within Christianity meant that, by the time of Cappadocia's rock-cut churches, the region was an ideal place for cosmic information to be hidden within church architecture, based on information not found in manuscripts. In such times, a persecuted tradition follows the path of survival through direct transmission of their secrets within enduring objects. This probably explains why the design of

---

*The controversy of the Iconoclasm spanned roughly a century, during the years 726 to 787 and 815 to 843, after the early rock-cut churches and before the unique art was painted in them by wealthy sponsors.

these churches was cosmological. The Orthodox style of churches in the Eastern Church evolved into the basilica, or "cross-in-square," which express the number 3 squared, and a central dome with up to eight peripheral directions. At least, this is my explanation for how Christians might have come to build the rock-cut cosmograms and why their iconography and liturgical arrangements were different from the norms of the early Western Church. Instead, churches were made to correspond with cosmological facts unknown to the outer culture of either the Roman or Orthodox Church.

The practice of building cosmically inspired buildings belongs to a "middle," or mesoteric, world in which "anything can be calculated"[2] using sacred numbers and geometry as these apply to the our geocentric world. Gurdjieff was giving a clue to this domain of "mesotericism," where what he called "legominisms" had been made to hide the profound within the design of a culture's ordinary objects, such as churches, art, ritual dances, games, music, and so on. He claimed an esoteric person would "read" these directly, like a book, but the mesoteric person must interpret such designs using the same design language that had been employed to make the legominisms, to serve the esoteric work of preservation.

An unfinished book by Gurdjieff's student J. G. Bennett[3] plotted the deep influence of northern Sufism upon world history, around the time of the thirteenth-century Mongol invasions of Europe and Central Asia. Its proposed ending was to describe the similar work of such masters in Europe, in European history. Only a footnote in Bennett's booklet *Creation* passes on Gurdjieff's hint that the Cappadocian liturgy held a wealth of cosmic information, as we shall next see, with the help of recently available plans. Gurdjieff, a Greek-speaking Armenian, must have visited the Cappadocian churches. It is very likely no other chain of transmission for these secrets exists, yet the power of legominisms lies exactly in the possibility that someday someone will read them, according to the ancient number sciences used to construct them.

My *Sacred Geometry: Language of the Angels* presented some of J. G. Bennett's ideas about the intelligence attributed to the interval between the Sun, our star, and the Earth, a planet with life. Bennett wrote the following footnote in his booklet *Creation:*

> The liturgy was a very remarkable construction for which a group in Cappadocia was largely responsible. . . . Gurdjieff claimed that it was one of the most extraordinary legominisms in the world, and all the secrets of the universe are contained in it.[4]

Within a few centuries of the liturgy described in the Creed, Cappadocia would see the building of rock-cut churches with astronomical invariants built into their rectangular and circular designs, and for which no such similar record has so far been found, being more advanced than the megalithic models found from between 5000 to 3000 BCE.

INTERPRETATION OF AYVALI KELISE

Now armed with the idea that liturgy and Cappadocia were cosmologically significant, a plan was found of the Ayvali Kelise, or "Quince Church," in the valley of Gullu Dere between Goreme and Cavusin (see fig. 7.3).

The plan in figure 7.3 was numerically enlarged to give measurements in 12/77-foot subunits instead of meters. The meter scale on the plan precluded

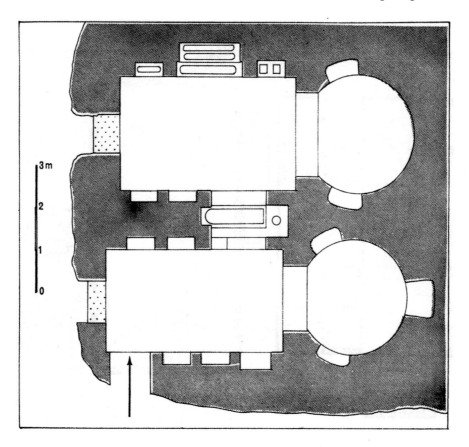

FIGURE 7.3. Plan of Ayvali Kelise. Found in Teteriatnikov, *The Liturgical Planning of Byzantine Churches in Cappadocia,* after Thierry, with permission of the publisher.

direct sight of foot-based subunits, metric measurement being unhelpful in the task of identifying ancient units of measure, before the modern meter was defined.*

An apse is circular, and a naos, or nave, rectangular. Both these spaces appear to employ whole numbers of Göbekli units in their principal dimensions (see fig. 7.4). Furthermore, the northern Ayvali apse has two symmetrical alcove seats whose depth define the circle of equal perimeter to the square, whose perimeter is four times the diameter of the inner apse: a relationship presented as the out-square of the apse, all shown in red. The out-square of the apse then has the naos placed where the concentric circle of equal *area* to the out-square would just touch the threshold to the apse, from the naos. The radius of this

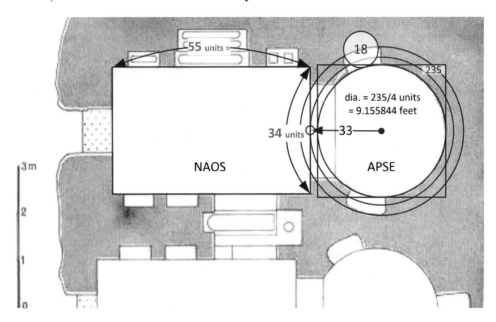

FIGURE 7.4. Analysis of the plan reveals the apse and naos conform to use of the Göbekli unit of 12/77 feet and 20/77 feet, respectively. The alcoves imply the equal perimeter model (radius 33, yet the apse wall, at radius 29⅜ units implied a new and unknown geometry of a 3-4-5 rectangle, with 235 units as the circumference (see fig. 7.5).

---

*Converting a plan's meter scale to feet makes the whole system of metrology, rational to the English foot, available through division and multiplication using a single calculation, and using a scale in 12/77-foot subunits saves time when testing for the possible use of those smaller units; that is, one does not need to convert each measurement from meters to feet and then divide that by 12/77, for each measurement.

(black) circle, from the center of the apse, is 33 units, giving it a diameter of 66, which is normally the inner circle of the geometry. We will discuss this later, as it contains a highly significant extra meaning beyond those seen before within ancient monuments.

Finally, the naos rectangle dimensions of 55 units long and 34 units wide reveal a golden rectangle whose sides express an approximation to the golden mean using the adjacent Fibonacci numbers 55 and 34, a ratio of 55/34 = 1.6176, just less than the golden mean itself (which is 1.618034 to six places). A golden rectangle drawn within the apse of Ayvali Kelise brought the recognition that the naos was a golden mean rectangle, and a smaller golden mean rectangle could always be drawn to fit inside any circle, including the apse.*

Of course, all possible *ratios* of rectangle can and will fit within any circle, and this freed up the thinking here. It showed that the Ayvali apse circle, of radius 29⅜ units, could form a 3-4-5 triangle in which the 5 side (the diameter) would be 235/4 units (9.155844 feet), so that the circle's out-square must have a perimeter of 235 units (see fig. 7.5).

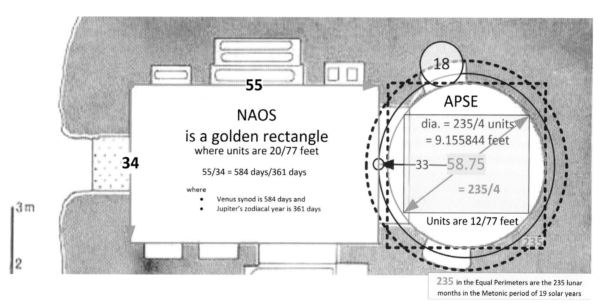

FIGURE 7.5. Ayvali Kelise, measured in Göbekli units, reveals the church's numerosity, meaning "the condition of being numerous (countable)" of three cosmic periods of 19 years in the apse of 235 lunar months (the Metonic period); 1.6 years (Venus synod); and 361 days, Jupiter's zodiacal year.

---

*Based on various considerations of Dan Palmateer of Nova Scotia.

For the apse, as an equal-perimeter circle, these 235 units pass through those sitting on the niche benches and those standing on the threshold. The astronomical significance of the number 235 is as the number of lunar months in the recurrence of Sun and Moon (and hence of its phase) relative to the zodiac; continuously repeating the full ensemble of possible configurations, over almost exactly 19 years, to connect the present to a moment 19 years ago and to the next such moment, 19 years in the future.* In the Western Church, the Metonic period allowed, after the synod of Whitby, mobility for the date of Easter, to depend upon the first Sunday after the first full Moon after the spring equinox. This was around the date at which these single-aisle churches were cut out.

These early single-aisle churches appear not to function as community churches but rather as private chapels, tended by a prominent monk and therefore termed a hermitage, and this one belongs to the second of two phases of rock-cut churches that have been called Early Christian & Iconoclastic (550–850), and then Archaic,[5] the latter of which being where most catalogues then place the construction of Ayvali Kelise, based on its later art rather than concrete criteria. I believe it to be Early Christian due to its cosmological content, linking it to Haçlı Kelise (below), which Kostof locates as Early Christian.

INTERPRETATION OF HAÇLI KELISE

Another single-aisle church was Haçlı Kelise, or "Church of the Cross," from the Early Christian phase (see fig. 7.6). Only discovered in 1963, it features an enormous cross on the ceiling and troglodyte furniture in lieu of the wooden equivalents found in contemporary stone-built churches. The apse is again circular but has a horseshoe-shaped step bench called a *synthronon*. Such seating is an unusual feature since elsewhere congregations would stand. And this apse allowed a select gathering on a horseshoe step bench, with a central *cathedra,* or bishop's chair, shown in the schematic plan in figure 7.7 on page 154.

The naos has the characteristic northeastern prothesis[†] niche, unique to early Cappadocian churches. The paintings would have been added after the

---

*In celestial dynamics this is known to be possible according to the Mirror Theorum or Caledonian Symmetric rediscovered after Poincare in late nineteenth century and by Roy and Ovendon in 1955, see *Orbital Motion* by A. E. Roy (Bristol & Philadelphia: IOP Pub 1978–2005).

†A prothesis niche is where the wine and bread of communion are placed, also called Holy Gifts for the divine mystery of transubstantiation, giving churches the form of the Last Supper of Christ.

FIGURE 7.6. *Top:* the volcanic cones with Haçlı Kelise on the right, in the valley of the Kizil Cukur. *Below:* the apse from the naos of Haçlı Kelise (Church of the Cross). Note the twelfth-century paintings and *synthronon* bench, with bishop's *cathedra* throne.
Photo by Jason Borges for the *Cappadocia History* website.

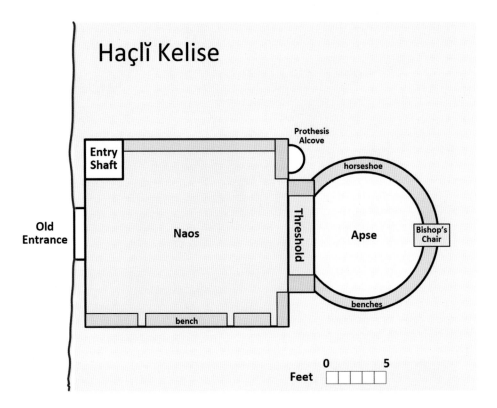

FIGURE 7.7. Plan of Haçlı Kelise, adapted from Kostof, *Caves of God.*

Iconoclasm, but crosses were not considered art and were a mark of a church being consecrated after construction (see fig. 7.6, *bottom*).

This church again has dimensions coherent with the Göbekli unit of 12/77 feet within its geometry. Looking at figure 7.8, the *synthronon* appears to have the same inner radius as the distance of Ayvali Kelise's threshold to the apse center (33 units), while the new threshold is now 42 units. So, that is 42/33, which equals 14/11—the ratio of the two circles in the equal-perimeter model, shown in blue. The apse wall appears just smaller than this: perhaps it was not so important since the square and circle simply equal 4×66 (264 units) in this equal-perimeter geometry, instead of the 235 units of Ayvali Kelise, deriving from 33 as the outer equal-area model.

Links to the Calculation of Easter

The word *synthronon* (*syn,* meaning "together," and *thronon,* "a throne") is a bench reserved for the clergy with the bishop's throne in the middle. A bishop's

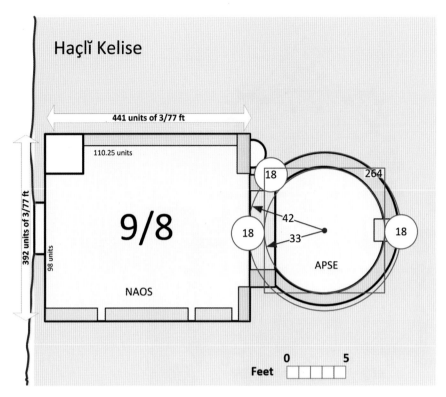

FIGURE 7.8. Analysis of Haçlı Kelise using Göbekli units or ¼ of these to achieve integer values. Plan from Kostof, *Caves of God.*

convention was called a *synod,* as in the Synod of Whitby (in northeastern England), which determined the way in which Easter Sunday was to be calculated. To establish Easter, 235 was broken into a number of short lunar years of 12 lunar months and longer ones of 13 lunar months using "golden numbers" to find the first Sunday after the next full Moon after the spring equinox.

It is relevant that

- 235 is 7 more than 228 (which is 19 lunar years of 12 lunar months), making the Metonic period of 235 lunar months rational, 7 extra months then equals 19 solar years;
- adding a single longer lunar year (of 13 months) every three lunar years, the lengthened lunar years will be 37 (12 + 12 + 13) months instead of 36; and
- these can be distributed every third year, so that 6 × 37 = 222 months, which one further longer year of 13 months achieves the Metonic; since 222 + 13 = 235, the Metonic period.

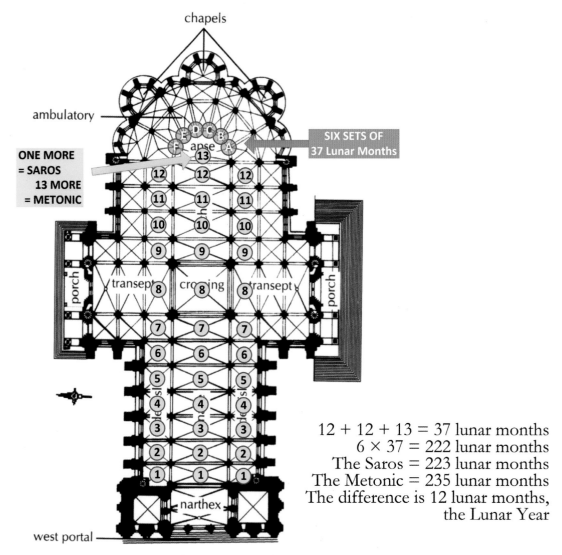

chapels

ambulatory

ONE MORE
= SAROS
13 MORE
= METONIC

SIX SETS OF
37 Lunar Months

E D C B
F apse A
(13)
(12) (12) (12)
(11) (11) (11)
(10) (10) (10)
(9) (9) (9)
transept (8) cro ing (8) transept
porch   porch
(7) (7) (7)
(6) (6) (6)
(5) (5) (5)
(4) (4) (4)
(3) (3) (3)
(2) (2) (2)
(1) (1) (1)
narthex
west portal

12 + 12 + 13 = 37 lunar months
6 × 37 = 222 lunar months
The Saros = 223 lunar months
The Metonic = 235 lunar months
The difference is 12 lunar months,
the Lunar Year

FIGURE 7.9. Plan of the Gothic cathedral of Chartres showing counting the
Metonic period to establish Easter.

The rectilinear form of the Gothic cathedral, punctuated with pillars and
pillared bays, itself extended the form of the early Greek temples such as the
first Heraion (Greek being one of the meanings of "Gothic.") The ability of
number symbolism and metrology to express cosmic time periods seems to have
continued into the medieval period, for the builders of Chartres (see fig. 7.9).
There were 12 bays in three central columns, to which the apse then adds one
more, to make a count of 37, or 12 + 12 + 13, lunar months.

To calculate Easter one needed to keep a running count of lunar months over the Metonic period of 235 as follows: 6 × 37 = 222 + 13 = 235 (the central column including the apse). The Saros eclipse period could also be known within this count since 222 + 1 (the apse) was equal to 223 lunar months, which are 19 eclipse years. The eclipse was associated with death, and the sacrifice at spring equinox was moved to the next Sunday after the full Moon. Good Friday therefore had a double delay, one based on the full Moon and the other due to which day of the week the full Moon was on. This suggests the work in Cappadocia went forward to reappear in Whitby and then the Gothic school, along with the week of seven days.

This echo of megalithic style calculations using lengths of time, occurring around the same time as the rock-cut churches, reveals a relationship between an inner circle practicing megalithic skills that could confound the previous fixed date of Easter (the traditional death of the king in a Saturnian year of 364 days plus 1). It was the megalithic that rationalized the Sun, the Moon, the Saros eclipse period, the Metonic period, and the seven-day week, but with their own religious ideas in play.

No wonder, then, that when the disciples met Jesus after his resurrection, having lived out the 33 years of the solar hero, he was to define the soli-lunar calendar triangle of 3-12-$\sqrt{153}$ with a net (made of squares) thrown on the "right side" of the right-triangle-shaped boat, to catch 153 fish, the square of 12 and 7/19 lunar months. Over 19 years this gave the 7 extra lunar months mentioned above. The boat may have been symbolic of the triangle, whose side was 12 units long and stern 3 units long. The number 153 equals the sum of the squares 144 and 9, as Pythagoras's theorem would have said six hundred years before.

The day of spring equinox was established using a marker from a backsight, just as the megalithic did. The next full Moon was shown on the lunar count in days from which the day of the week gave the next Sunday, which was now the "Lord's Day," after the "Sabbath" and before that Friday (of Venus) became the Muslim equivalent, as People of the Book. This clever legominism forced these skills to be valued while showing Jesus to be a solar hero and a "cosmocrator" like his father, who created the world in a week.

The number 19 is called the Signature of Allah in Islam, a cosmocratic role Jesus inherited as son of God from his father. But 19 is also a number most clearly articulated by Jupiter, rather than Saturn, who *handed over the measures* of rulership to Zeus, as seen in the trigon calibration of the precession of the equinoxes, see figure 8.3 on page 178.

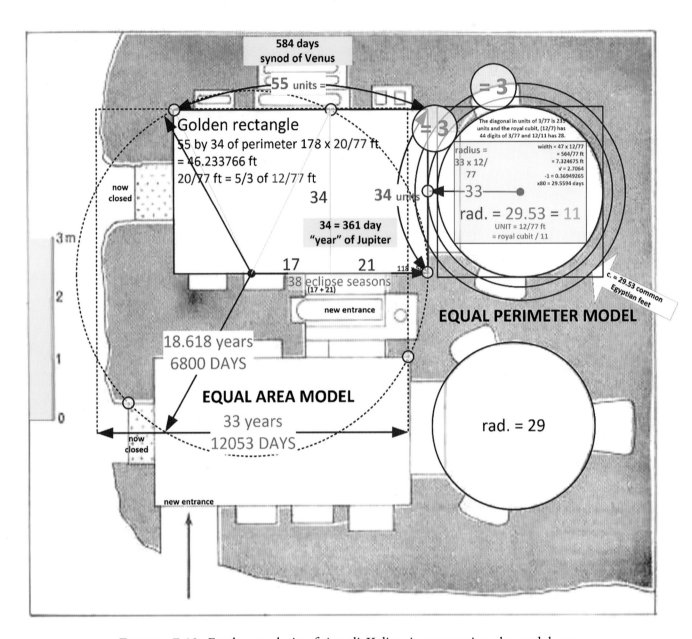

FIGURE 7.10. Further analysis of Ayvali Kelise, incorporating the nodal year of 18.618 years and the solar hero period between the north and south naos, and the Metonic period in the 3-4-5 extension of the apse 235/4 in diameter, something forgotten in our own equal-perimeter lore. The diagonal in units of 3/77 is 235 units, and the Royal cubit (12/7) has 44 digits of 3/77 and 12/11 has 28. This means 132 x 1/77 in Royal cubit and 84 in Sumerian foot. (See also the discussion in chapter 5, "The Monolith Basement at Knossos.")

The face of Jesus in art had become fully Saturnine by 600 CE,* and Saturn's name in Greek, Cronos, equated to Chronos, the god of time and hence Saturn by the Neoplatonist Microbius. The Christ was equated to the new Aion (Age) of Pisces as Pantocrator, Lord of the World. This Saturnine appearance (thin faced and bearded) was shared by saints and monks and was illustrated in their faces, which are serious and with mouths downturned, showing their work of penitence. The presence of high *numerical art* in the rock-cut churches, close to where the liturgy was defined by the three Cappadocian fathers, after the Roman Church had relocated to Constantinople by Constantine, suggests hidden influences were at work, perhaps masters of wisdom to help Christianity and who, while guiding history, left their surprising telltale signs within such legominisms.

EVOLUTION OF ORTHODOX CHURCHES IN CAPPADOCIA

In the sixth to seventh centuries the rock-cut single-aisle church (see fig. 7.11 on p. 160, *top left*), of nave plus apse and other rooms, became two parallel aisles (*top right*), with this design allowing a space to hold mundane services such as funerals and a Sacramental space for other rites. Then a third design, the triple-aisle basilica church came into existence, the norm for Greek Orthodox churches. What had started off as two or three distinct aisles was inflexible. Dividing a square into nine smaller but equal squares, the cross-in-square basilica church symbolized the spiritual Earth as a central square within the prevailing geocentric planetary model thought to be Chaldean in origin (fig. 7.11, *bottom left*). At its most simple, a square of side length 3 has an area of 9, and that number matches the geocentric planetary model, with the Earth, its Moon, six visible planets, and the zodiac, in a Chaldean model that survived until the sixteenth-century heliocentric model.

The Equal Division of Squares
There are two ways to divide any square into a whole number of smaller squares of equal size, with the square's side length being either an odd or an even

---

*Allen, in *Early Christian Symbolism in Great Britain and Ireland,* also noted that Saturn appears as a controlling figure in Hesiod's *Theogony* when, I presume, the matriarchal cultures of the Mediterranean are being characterized. Hence the "mother and child" in the story of Jesus already has the primacy of Mary to have the son of God, the latter being bonded to both matriarchy and patriarchy through being of the line of David, the patriarchal king.

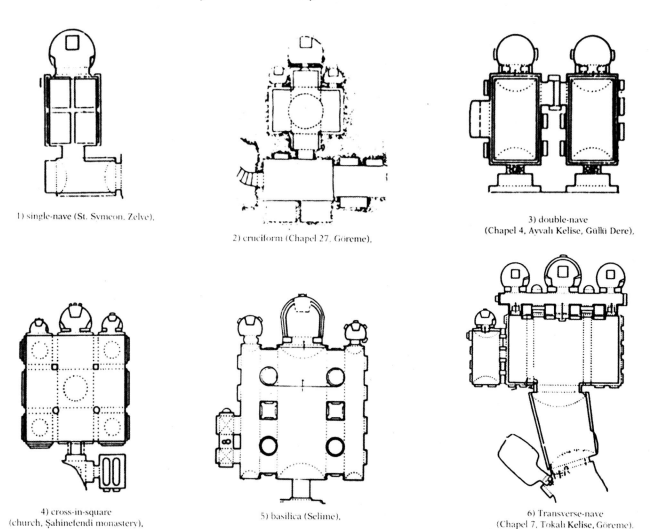

1) single-nave (St. Symeon, Zelve).

2) cruciform (Chapel 27, Göreme).

3) double-nave
(Chapel 4, Ayvalı Kelise, Güllü Dere).

4) cross-in-square
(church, Şahinefendi monastery).

5) basilica (Selime).

6) Transverse-nave
(Chapel 7, Tokalı Kelise, Göreme).

FIGURE 7.11. The evolution of rock-cut Orthodox churches from single-aisle
to cross-in-square. Diagram from Teteriatnikov, *The Liturgical Planning of
Byzantine Churches in Cappadocia.* Figure 1, after Rodley (1, 3–6),
Kostof (2), with permission of the publisher.

number of building units. The smallest such square is side length 2 (area then 4)
and every even-sided square after that {6, 8, 10, etc.} will have a 4-square at
its center. In contrast, odd numbers as side lengths always have a single square
at their center, starting with side length 3 (see fig. 7.12), if you discount a side
length of 1 as just simply a square.

It is natural to visualize squared numbers geometrically as squares, hence
the name square for us, but (fishing) net, (weaving) mesh, etc., are metaphorical

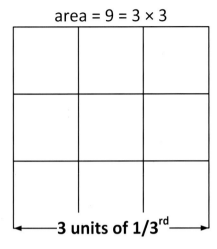

area = 9 = 3 × 3

◄——— 3 units of 1/3rd ———►

ninefold triple-aisled church

FIGURE 7.12. The division of the square triple-aisle church by a grid of nine equal sub-squares. There is a central square whenever the side length is odd.

early names in other languages. The division of geometrical squares was easy to study in antiquity or the Stone Age, to see the two types of square as odd and even.

The ninefold square allowed for one aisle, the central square, to be used by congregations as a naos for communion and other services such as baptisms and marriages. Funerals, focused in other directions, used the outer squares. Alternatively, the Orthodox veneration of icons could create spaces of devotion with gallery spaces for the icons, as in Western cathedrals. We will see that the square design is eminently suitable as a sacred space in relation to ancient cosmological models such as equal perimeter. Such basilicas often have a dome, representing the Northern Hemisphere, [6] above the central square. The diameter of this dome is then the central square's side length × $\sqrt{2}$. This allowed the dome to be supported by columns on four corners, by arches to distribute the weight of a dome. In rock-cut churches where the roof was part of the bedrock, false supports were used to recreate this design. The theological issue of overusing a single aisle as communion altar and other purposes was thus overcome. A centralized congregation within the square could access the eight outer squares of the naos. Troglodytic benches and other furniture seen in the single-aisle hermitage churches were probably reserved for special meetings of the clergy. The central dome was the inheritor of the apse, and the central square made larger, at the expense of the outer squares, to provide an enlarged naos surrounded by eight different directions (see fig. 7.13 on p. 162).

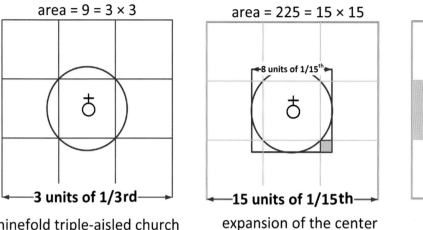
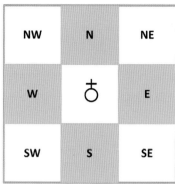

area = 9 = 3 × 3

3 units of 1/3rd

ninefold triple-aisled church

area = 225 = 15 × 15

8 units of 1/15th

15 units of 1/15th

expansion of the center

NW	N	NE
W	♄	E
SW	S	SE

its implied cardinality

FIGURE 7.13. The supported dome and its expansion within a structure naturally cardinal to the compass directions.

Numerical Properties of the Ninefold Square

Another feature of the 9-square is that the first nine numbers {1, 2, 3, 4, 5, 6, 7, 8, 9} can be located in the square with 5 as the center so that the number pairs equidistant from 5* and placed opposite each other create a so-called magic square in which all the rows, columns, and diagonals add up to 15 (see fig. 7.14). This ninefold square was named by the Medieval period after the planet Saturn, and it is a strange fact, as mentioned above, that the image of Jesus Christ changed from being like the Sun, or Apollo, into the Saturnine form by the time of Orthodox religious art so that domes and apses often had art depicting a Saturnine son of God who then matched his father, as bearded. (The squares with higher numbers of squares in their sides were named for other planets.†)

Note that all the numbers, in all the rows or all the columns, sum to 45, the number of Adam. This means the first nine numbers sum to 45, which is the square of 3 (9) times 5, the central number in the set {1, 2, 3, 4, 5, 6, 7, 8, 9}. Each of the numbers is used twice: for example, 4 appears in the sums (4 + 9 + 2) and (4 + 3 + 8). Abraham's wife, Sarah, conceived Isaac at 90 years of age, the total of all the sums of rows and columns, and Moses, who died at 120, has the further diagonals of (15 + 15) equaling 30 added. Isaac died at 180 years so

---

*{1, 9} {2, 8} {3, 7} {4, 6} all adding up to 10.
†4 Jupiter, 5 Mars, 6 the Sun, 7 Venus, 8 Mercury, 9 the Moon.

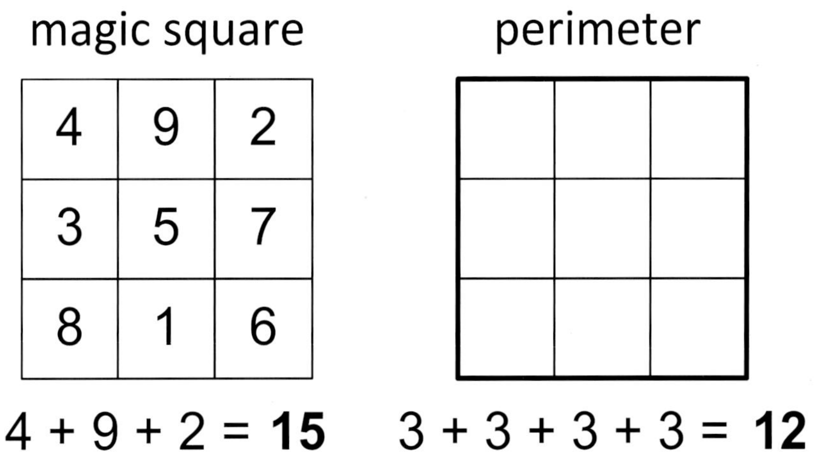

magic square

4	9	2
3	5	7
8	1	6

perimeter

$$4 + 9 + 2 = \mathbf{15} \qquad 3 + 3 + 3 + 3 = \mathbf{12}$$

FIGURE 7.14. The magic square of Saturn and perimeter of the 3-square.

that the sums of the columns and rows seen from the four cardinal directions could represent Isaac, who was nearly sacrificed by his father, Abraham. And if one divides the unit by 16, the 45 becomes 720, the limiting number for the onset of just intonation, born of just the primes {2, 3, 5} and being factorial 6 or $1 \times 2 \times 3 \times 4 \times 5 \times 6 = 720$.

The perimeter of the square in units is 12, the number of tribes of Israel but also the number of the zodiac's constellations. Twelve feet (11 Sumerian feet) are $12 \times 77/12$ (77) Göbekli units. Eleven is the signature diameter of the equal-perimeter model (see fig. 1.4), which is therefore the natural unit, as found at Westminster Abbey's coronation pavement (see fig. 7.18 on p. 168, *top*). If one scales up by 4 English feet to each foot of the 12-foot perimeter, then the out-square equal in perimeter to the square is 4×11 (= 44) Sumerian feet, as would be the case of the equal-perimeter circle of diameter 14, when π is assumed to be 22/7—this is the "magic" of this geometry.

The number 11 brings us directly into contact with the model of equal perimeter in which the new perimeter of 44 Sumerian feet will have a circle of equal perimeter whose diameter is 11 Sumerian feet times 14/11 feet, which is 14 Sumerian feet, radius 7; that is, the ninefold cross-and-square churches, in their simplest form, gave the primary form of the model when using 11 Sumerian feet, or multiples of that length, to build the outer square of the church. It is possible that the Westminster pavement embodied this.

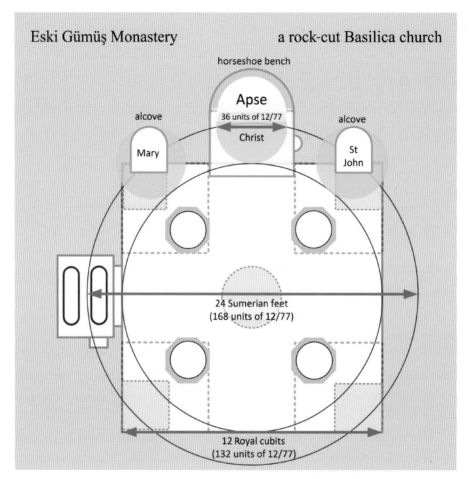

FIGURE 7.15. The equal-perimeter model drawn over the Nigde, Eski Gümüş room 19 plan. After Rodley, *Cave Monasteries of Byzantine Cappadocia.*

ORTHODOX TRANSMISSION OF THE PERIMETER AND AREA MODELS

If we now turn to a basilica-style rock-cut church of probably the tenth century, the monastery of Eski Gümüş (see fig. 7.15), we can see it using the ninefold design with 4 Royal cubits as the modular side length, and hence there were 12 royal cubits across the full square. Remembering that each Royal cubit is 11 subunits (of the style found at Göbekli Tepe and in Malta), then the sides of this basilica are 4 cubits × 11 each × 3 cells, or 132 subunits. This square is guaranteed to equal the perimeter of a circle with a diameter of 168 subunits (132 × 14/11) within the equal-perimeter model. The in-circle to the square

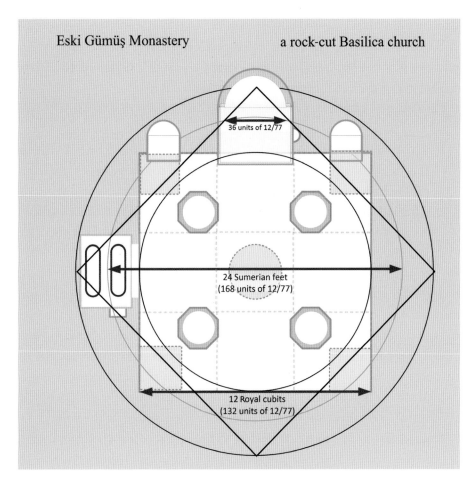

FIGURE 7.16. Eski Gümüş basilica with a blue circle for the Jupiter synod of 9/8 relative to the equal-perimeter circle, and the Saturn synod (black) as a diamond of 16/15 of the equal-perimeter (rock-cut) square.

will symbolize the size of the mean Earth. Multiplying 132/11 (12) by 3 obtains the diameter of the Moon in the model as 36 units, drawn in blue.

In figure 7.16, note the presence again of the Göbekli subunit. The Royal cubit of 12/7 feet occurs naturally as 11 subunits of 12/77 feet, and the subunit enables the construction of smaller versions of the equal-perimeter model, allowing it to be realized within a rock-cut square of side length 144/7 (20 + 4/7) feet. But that square only affords the out-square within which the mean Earth circle sits, but also present are planetary synods of outer planets as seen in chapter 7, "Pavements of the Savior," in my *Sacred Geometry: Language of the Angels.*

The design of the basilica manages to escape the confines of the rock-cut square by using the elongated apse to sample larger concentric geometries, with a circle of radius larger than the equal-perimeter circle by the ratio 9/8, to model the synod of Jupiter. Saturn was captured using a diamond 16/15 times the diameter of the equal-perimeter square (see fig. 7.16).

The two new geometries enlarge existing dimensions in two ways:

1. The radius of the equal-perimeter circle was enlarged by 9/8, a length that reaches into the apse as an arc but is too large to fit the equal-perimeter square of the church
2. The half-diagonal length of the equal-perimeter square was enlarged by 16/15, to form a half diagonal of a diamond (Saturn) that then, projected into the apse from the center, just exceeds Jupiter's radius

Through this process, focused on the apse, Saturn is shown superior to Jupiter. Saturn's corner delimits the benches of the apse, and the diamond then contains the communion niche. Jupiter's circle is outside the church except where its radius transits the apse. If geometry can carry symbolic meaning, the esoteric doctrine corresponds to the development from the sixth-century onward in which Christ had become Saturnian, and the king of the gods Zeus-Jupiter was relegated and Christ substituted as cosmocrator or lord of the world.

The fourth-century formulation of a defining Creed for the church by the Cappadocian masters was designed to identify the many forms of heresy that early Christianity had adopted, including the Gnostic and theological cults that were prevailing in many regions. Even when Saint Basil founded the work leading the Creed of both Rome and Constantinople, his home region of Cappadocia was in the grip of Arianism, which questioned the fully divine nature of Jesus. The Persians and then Islam were turning Cappadocia into a military frontier, but largely as a route for armies passing by the difficult valleys of the rock-cut churches.

The Saturnian face of Christ has strange resonances, some probably accidental. Like the Creed itself, it was effective in the difficulties of the centuries ahead. The literature of the Greco-Roman world would reach medieval Europe through being preserved in monasteries and also coming to light through contact with Islam's own renaissance of learning around the time of the Crusades, which resulted in texts, lost to the monasteries, being translated from Arabic into Latin. Realistically, therefore, the Mediterranean's past astronomical knowledge at this time was going underground into those forms Gurdjieff later described as legominisms.

THE CAPPADOCIAN CROSS

In chapter 7 of *Sacred Geometry: Language of the Angels,* I describe two English church pavements, one in Westminster Abbey and another in Canterbury cathedral (see fig. 7.17). Both can be developed from the equal-perimeter model, and the latter clearly (and concentrically) also expresses the equal-area model, a model that describes the motion of the Moon's orbital nodes, the synods of Jupiter and Saturn, and the solar hero period of thirty-three solar years.

In the top right of the Westminster pavement (fig. 7.17, *left*), the north-eastern roundel has the iconic Cappadocian cross design. This implies that the Cosmati school, credited with building the Westminster pavement, had definite links to Cappadocia. The Canterbury pavement further added the astronomy of the lunar nodes and the solar hero, through the equal-area geometry, with the ratios of Jupiter (the outer square) and Saturn (the outer diamond), to the lunar year (see fig. 7.18 on p. 168, *bottom*).

The geometry of the Cappadocian cross follows some general constructional rules. The negative spaces of the background circle can be defined by two 3-4-5 rectangles at right angles. The cardinal points of the circle then define a radius to the corners of the 3-4-5 rectangles to define the arced sides

FIGURE 7.17. *Left:* the Cosmati pavement at Westminster Abbey. *Right:* the mosaic pavement before the shrine of Thomas A. Becket in the Cathedral Church at Canterbury.

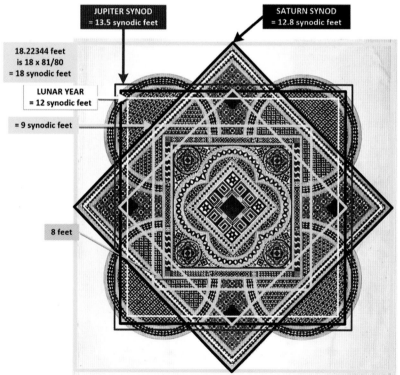

FIGURE 7.18. *Top:* Westminster Abbey's Cosmati pavement completed in 1268 CE. *Below:* Canterbury's pavement installed ca. 1120 CE. Both use the foundational square to derive circles of equal area and perimeter, with related astronomical meanings. From my *Sacred Geometry: Language of the Angels,* chapter 7.

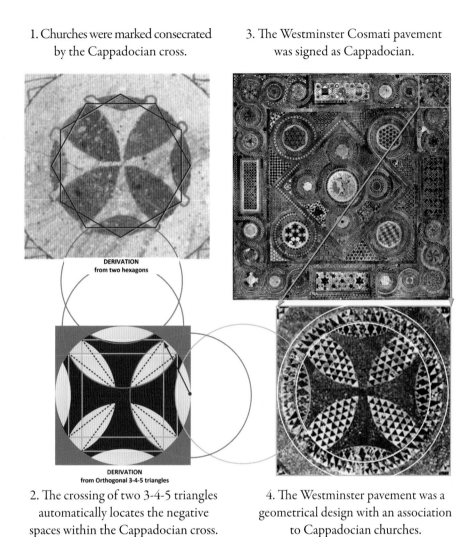

1. Churches were marked consecrated by the Cappadocian cross.

DERIVATION
from two hexagons

DERIVATION
from Orthogonal 3-4-5 triangles

2. The crossing of two 3-4-5 triangles automatically locates the negative spaces within the Cappadocian cross.

3. The Westminster Cosmati pavement was signed as Cappadocian.

4. The Westminster pavement was a geometrical design with an association to Cappadocian churches.

FIGURE 7.19. Iconic crosses of Cappadocia with eight points on their tips. *Top left:* using a double hexagon to define the tips; *bottom left:* using two {3-4-5} rectangles to define them; *bottom right:* that method used in the Cosmati Pavement; and *top right:* the example found on that Westminster pavement in situ, in the northeast corner.

of the four crosses (see fig. 7.19). A suitable arced radius is then used to give a thickness to both the tips of the cross and the size. This appears to lead to the version found on the Westminster pavement. In practice the crosses were probably drawn within the churches using two hexagons pointed at zero and thirty degrees to define the points of the cross.

8

THE ARCTIC ORIGINS OF
ASTRONOMICAL SYMBOLISM

The connection of the Earth to the bright stars, planets, Moon, and Sun is expressed in a remarkable way at the North Pole, and, during the Late Glacial interstadial (12,700–10,800 BCE), or "interglacial period," of nearly two thousand years, the warm current of the Gulf Stream gained access to the Arctic Sea. The partially unfrozen Arctic enabled human habitation near the European Arctic coast while most of Europe was covered by a sheet of ice kilometers thick* and humans were forced to live at the lower, warmer latitudes. But the end of the Ice Age was precarious. The meltwaters caused by the interglacial period closed down the Gulf Stream, causing a very cold period called the Younger Dryas between 10,800 and 9,600 BCE. "By 9600 BC the cold spell was over and by 8000 BC, temperatures had reached much the same range as they are today"[1]

One hundred years ago an Indian scholar named B. G. Tilak realized that if humans could have lived near the North Pole during the interglacial period before 11,000 BCE, the gods and demons of the Vedas, perhaps the oldest oral texts in the Indo-European language, would make more sense if they had been inspired by the skies within the Arctic Circle beyond 66 degrees north in latitude, and especially on landscape where the Arctic Ocean would have been up to 100 meters lower, due to the water retained by the ice sheets. For instance, the obliquity of the Earth, the tilt of the planet's rotation relative to the plane of the solar system, along which the Moon and planets travel, hides the Sun during

---

*The water trapped in glacial ice is largely concentrated from the compaction of snow, especially in glaciers formed by continental mountains.

the winter when it travels along its southern half. At the pole itself, this means there is a yearly rather than daily sequence of Day and Night, henceforth capitalized. At the pole, there was a single Day and Night every solar year, which corresponds to what it says in the Vedas—namely, that the year is a single Day for the gods and that the pole is its greatest hero, Indra, who defeats the dragon, thereby raising the Sun above the polar horizon.

If so, the Vedas were the result of astronomical work to describe what was seen, using a spoken-verse form then carried through time by the institution of priests, originally termed *rishis,* and subsequently, in India, by the Brahmin class until the oral tradition was phoneticized into sacred texts as an alphabetic language called Sanskrit,* but still also memorized by the priests.

At the pole, the rishis saw the sky in a unique and simpler form than seen from any other latitude on Earth. This form is less clearcut in the megalithic stone circle designs of Britain or Brittany, where the whole of the Sun's path can be seen, even in the dark half of the solar year. Latitude raises the celestial equator above the horizon as one moves south. On the Arctic Circle, the place of the midwinter Sun is just visible all Day, so that conventional days and nights ensue; instead of one Day and Night at the pole, 365 normal days appear per year.

It is important to stress that the polar Night is often not fully dark since there are the northern lights and the Sun creates extended periods of dawn and dusk: "Astronomical twilight happens when the Sun is between twelve and eighteen degrees below the horizon and astronomical night when it is lower than that."[2] At the actual North Pole the celestial equator has sunk down to the horizon, so that only the northern half the Sun's path is visible by Day. Dusk, Night, and Dawn last half a year, when the planets and stars can be seen as a "Vision in the Long Darkness,"[3] the title of a significant poem in the Rg Veda. Day lasts six months, with the Sun steadily climbing, all the while rotating "west" as usual to describe an all-around helix due to the Earth's rotation, until it reaches apogee at summer solstice. The Sun was therefore seen as climbing a holy mountain, then descending to the autumnal equinox, and then falling into the underworld ruled by a dragon called Vritra, who *was* the horizon as well as the celestial equator: they were conjunct. To reach the actual North Pole, there may still have been ice, but it is also possible to use one's imagination to see

*Phonetic alphabets developed in India during a period called Sanskritization, as in Greece, and not earlier than the first millennium BCE. All the religious texts of India adopted Sanskrit, even the Tantras of the Shakti (that is, Goddess) scriptures.

what must be the case at the pole, and a mixture of the two once living close to the pole, whose power over astronomy becomes palpable.

As we see, evaluating whether the Rg Veda was inspired by the astronomy of the Arctic requires an understanding of the form the Arctic sky presents, and of how the Vedic gods and demons were matched to these phenomena, requiring the sort of descriptive language capable of elucidating the details; that is, a work of both astronomy and linguistic innovation appears to have run in parallel over at least a thousand years. If so, then the Arctic was an important influence in both the formation of the Proto-Indo-European (called PIE), from which the group of languages we now call Indo-European, developed. I learned of Tilak via J. G. Bennett's notion that languages can only develop in isolation from other groups and that for a descriptive language to evolve there needs to be a powerful phenomenon and the desire to describe it.

The verses of the Rg Veda were orally chanted by the rishi-astronomers as they speculated on the roles the cosmic entities of light play to explain the form of the celestial world as having emerged from the darkness below the horizon. This explains many disjointed beliefs such as that the Sun goes under the Earth and that new stars are continuously being born from the darkness of the spring equinox and die at the autumnal equinox as the pole (Indra) moves among the stars in Plato's Great Year of about 26,000 years.

THE PHENOMENOLOGY OF INDRA

Indra was born to the maker of heaven and earth at a time when the Adityas, the gods of light, were in a stalemate with the Danavas, ruled by Vritra, the dragon of darkness. The Northern Hemisphere of stars had been created by the eagle Varuna and the hidden Southern Hemisphere by Vritra, who is at the border of subliminality, the horizon that originally ran alongside the road of the Adityas that were "in his belly." Indra makes preparation for a great battle, drinking three beakers of soma (the food of the gods).

The Earth is rotating, but the celestial equator, ecliptic, and horizon are stuck together as Vritra. Indra asks Vishnu (the ecliptic pole) to get ready to take a giant step (the obliquity of the Earth). He then confronts Vritra, who breaks his jaw (the ecliptic), so Indra breaks Vritra's jaw (the celestial equator), and the two jaws then fly apart, exposing the northern arc of the ecliptic upon which the Adityas, including the Sun and planets, move, above the celestial equator.

Vishnu, as the ecliptic pole, has now taken a great step away from Indra, as the North Pole, which now has obliquity of around 24.4 degrees.

THE ARCTIC FRAMEWORK AND ITS MOTION

A human at the North Pole would experience a single Day and Night within the solar year: the Day would be six months long, with Dawn or Dusk taking 30 days,* which leaves a four-month period of Night. The following phenomena would be observable:

1. the northern half of the ecliptic would continuously rotate around the observer
2. the Sun and planets, by Day, would circle "west," each at the *same altitude* above the horizon, during each rotation of the Earth
3. the solar day-length of 24 hours would be replaced by the rotation of the Earth, or sidereal day, which is 4 minutes shorter than the solar day
4. the Zenith above would also be the North Pole, and in that epoch the polestar was Vega in the constellation Lyra
5. the equinoctial points of the Sun, at spring and autumn, would also rotate, but on the horizon at 0 altitude, again continuously rotating to the right during every Earth rotation

These features are shown in figure 8.1 on page 174.

The key insight is that, as ever, the ecliptic and celestial equator only touch at two equinoctial points—when the Sun is visible at sunrise and sunset. And the celestial equator, being on the horizon (point 3 above), make Dawn and Dusk synonymous with the spring and autumnal equinoxes, thus cutting the solar year into a single Day and single Night plus a protracted Dusk and Dawn where the Sun is only moving slowly, lower or higher, on the southern ecliptic, which is below the horizon.

In contrast, the familiar form of megalithic astronomy was dependent on the daily horizon events of Sun and Moon moving between the four points or pillars of the year: this rising and setting of the Sun on the horizon punctuates the year with a countable 365 dawns and dusks. But in the Arctic, the

---

*Note this as a possible source for the preference of 360 for the degrees in a circular perimeter, and in harmonic theory relating to the planets and the days in a year, 360 having many small divisors—namely, the prime numbers 2, 3, and 5.

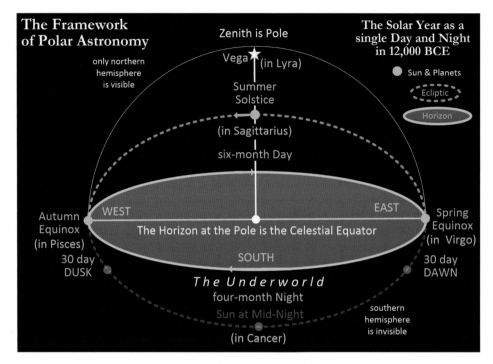

FIGURE 8.1. Some key features of the Arctic sky.

solar year calendar was hidden by the Day and Night, both being half a solar year long. Instead, as Day began, the whole disk of the Sun was seen on the horizon at spring, hiding the constellation standing at the spring equinoctial point, one of the four "pillars" of the year. The constellations above the rotating equinoctial pillars were clear to see at Night. During the interglacial period, circa 12,000 BCE, these pillars were Virgo (spring), Sagittarius (summer), Pisces (autumn), and Gemini (winter), opposite to those of our Current Era since 0 CE in the Age of Pisces (spring); that is, by the Christian (or current) era, the pillars are the same but have slowly moved on by half a turn (180 degrees) to be Virgo (autumn), Sagittarius (winter), Pisces (spring), and Gemini (summer). The Christians likened the current scenario with Jesus being the fish (Pisces) and his mother, the Virgin (Virgo) Mary, Jesus represented as cosmocrator (cosmic creator) of the geocentric worldview.

From moment to moment, therefore, all of these framework features, including the stars, rotated daily around the pole at the zenith, and, in 12,000 BCE, there was a polestar, Vega in the northern constellation of Lyra, directly above, at the polar zenith.

The Moon

Each orbit of the Moon (in 27⅓ days) takes it below the horizon (into the underworld of the Vedas) and then out again. What the Sun does in a Day and Night lasting a year, the Moon does in its orbit.

1. During the Arctic Day, the phase of the Moon is a new Moon as it passes the Sun (as usual), and during the Night, the full Moon would (as usual) be diametric to the solar orb's location in the underworld, in the southern underworld.
2. Therefore, at mid-Night, only the light half of the Moon is in the north, while the mid-Day noon would not show, being the dark side of the lunar month.
3. Lunar eclipses could therefore occur only at Night, when the Sun is hidden to the south and the Moon is illuminated. Solar eclipses by the Moon could only occur in the Day, when the Sun is visible. Only one lunar node can be in the northern sky: the ascending node where the Moon is crossing from south to north, or the descending node, when crossing north to south.

From this, one can see how our word *underworld* came to be associated with an invisible Southern Hemisphere that ate celestial objects, especially the Moon, in a different type of eclipsing of the Sun by the Earth. Objects below the horizon were lost, except by inference, when the Moon was being illuminated at Night by the hidden Sun. Celestial objects were being eaten and then reborn, and the Moon was the prime example. For these reasons, the Moon was quite crucial in giving a clear and natural order to the ecliptic.

The ecliptic can be divided into 137 parts since the Moon orbits the Earth in 27.4 (27.3908603) sidereal days, which, multiplied by 5, is 137 (136.98) sidereal days. A radius rope of 137 Göbekli subunits of 12/77 feet would create a circle whose perimeter would be 137 units of 48/49—the foot called common Egyptian encountered at Göbekli Tepe—and would make a circle calibrated to the Moon on the horizon. The Moon would then move five feet every day, in a simulation of the Moon's orbit in the (ideal) polar sky. This would be able to predict when the moon would rise and set while tracking the moon's orbital position in a concrete way. Such a circular "observatory" reminds us of the form of Göbekli Tepe's Enclosures that are around this size of 13 meters.

The maximum and minimum standstills of the Moon were particularly clear

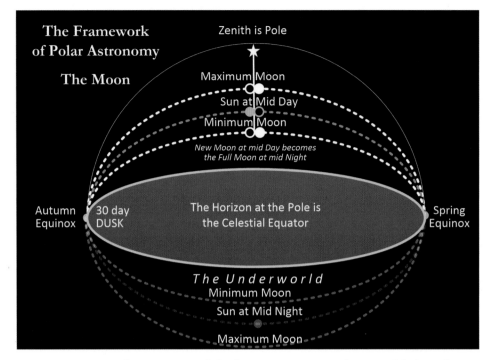

FIGURE 8.2. The extremes of the lunar maximums, above and below the ecliptic over 18.618 years.

at the Arctic, since moonrise and moonset would only sit on the equinoctial pillars when the lunar nodes were also on the equinoctial pillars. Otherwise, the Moon's rising and setting would be ahead or behind the rotating equinoctial points.

Jupiter, the Moon, and Its Nodal Cycle

The 3400 solar-day period between maximum and minimum Moon and the 6800 solar-day full nodal cycle would have to be achieved with reference to the sidereal day. In this respect, the synod of Jupiter, lasting 398.88 solar days, would be viewed as the 400 *sidereal* days between Jupiter's loops. Seventeen Jupiter synods would therefore be 6800 sidereal days. This corresponds to the Sumerian foot (SF/88) height of the equilateral triangle between enclosures B, C, and D at Göbekli Tepe, a triangle that points toward Vega, the polar vulture star in the constellation Lyra, also shown at Göbekli Tepe (see fig. 2.8). This allows that triangle to have 400 units of SF/8 vertically and 231 as its half side equaling $\sqrt{3}$ (= 400/231, as with the Preseli triangle discussed in "The Bluestone Culture" in chapter 3), only 17 has been removed because 6800 divided by 17 equals 400. The removal of 17 in the compact scaling at Göbekli Tepe can now be seen as an obvious step of scaling for the polar astronomers, for whom the sidereal day was primary and the solar day not available for counting between horizon astronomy events.

The Precession of the Equinoxes

The zodiac of twelve signs was a way to note the constellation obscured by the Sun at Dawn; that is, at spring equinox, which is slowly changing. The precession of the equinoxes (the movement of the Earth's equatorial nodes) is a much longer cycle than the Moon's nodal cycle. Taking around 25,920 years for the tilt of the Earth to complete its gyroscopic journey between occasional polestars and again, at the pole, during the interglacial period, it was more clearly seen. Vega was the polestar in the constellation Lyra. Most logically, the movement of the equinoxes is measured as the number of years required for them to move a day in angle among the stars, which takes just under 71 years, the traditional "four score and ten" span of a human life. It takes 18.618 years for the Moon's nodes to traverse the whole zodiac so that lunar maxima occur nearly 4 times in a "lifetime," and alignment to the maximum Moon, rising on the horizon, was quite universally thought of as having funerary significance.

As Fate, the ratio $\sqrt{\pi} = 1.77245$ contrasted with the solar hero period of 33 years as shown in the equal-area geometry. One is reminded of the greatest lyre player of all, Orpheus, who ventured into the underworld (to the south) to heal Eurydice, who, like Lot's wife in the Bible, against advice, looked back when leaving.

South of the Arctic, the celestial equator seems to have been given disproportionate significance in precessional myths, where stars are swallowed by the autumnal equinox or born at the spring equinox. Only in the Arctic will a star truly disappear below the celestial equator, and so be lost for millennia. When a star rises above the celestial equator this is resonant with resurrection of some god-power that had been lost in the deeps of the south. If this is only true at the North Pole, then the attitude expressed toward lost and resurrected stars, at lower latitudes within precessional myths that were first developed within the Arctic, likely means these tales are relics of polar astronomy.

Around 12,000 BCE, observers at the pole would see the movement of the North Pole, at the zenith, as small changes in the rotation of Vega around the North Pole as the Earth rotates. But at the two equinoctial points, rotating

The Equilateral Triangle is the √3 nodal cycle to Jupiter

The Arches are the Ecliptic at the Pole

Maximum Moon

Ecliptic

√3

DISH

POLE

Autumn Equinox

Spring Equinox

DOVE 17

MOON

VESICA

WAND

WORD

MARY

Gabriel is the Moon

The Dish is the 17 synods of Jupiter in the nodal period

FIGURE 8.3. Bernard van Orley's painting *The Annunciation* was a geometrical story that could have Arctic roots. As ever, the angel Gabriel is shown as an equal-perimeter Moon with Mary central, surrounded with symbols of her virtues. Gabriel's wand angles at 60 degrees and above one triangle is another with the dish of 17, which number is the diameter of the outer circle, in lowest numbers for equal perimeter—namely, {3, 11,14}. The equilaterals are then like those at Göbekli and Preseli. The arch can then be seen as an aspect of polar astronomy, for the maximum Moon, as we have described earlier, and the dish is of the 17 synodic loops of Jupiter within the nodal cycle of the Moon.

"west" of the horizon, existence (Sanskrit *sat*) was defined by which stars and planets were within in the Northern Hemisphere. Stars were becoming invisible at the spring equinox each Dawn and removed from existence at the autumnal equinox, during every Dusk. Whatever could not be seen was swallowed by Vritra, the dragon of Asat, or nonexistence.*

If the polar astronomers knew that 17 complete Jupiter synods (its loops in the sky), each of 400 sidereal days, defined the Moon's nodal period between maximum standstills (6800 of our solar days), then they could have seen that the loss of old stars and birth of new ones at the equinoxes revealed a larger period of the Great Year in which the stars at the polar zenith were also moving.

---

*The spiritual meaning of the RgVeda's cosmology is comprehensively given by Antonio de Nicolas in his *Meditations Through the RgVeda*, especially in chapters 4 and 5. De Nicolas obtained his astronomical facts from W. Norman Brown, page 121 section II. This completes my own three sources as de Nicolas, Brown and Tilak, though de Nicolas and Brown do not mention Tilak or the Arctic.

9

THE VEDAS IN
SOUTHEAST ASIA

Time creates the sky and the earth. Time creates that past and the
future. By Time the Sun burns, through Time all beings exist, in
Time the eyes see. Time is the lord of all.

ATHARVA-VEDA (19.54)

The previous chapter presented the case, made by others, that the earliest poems of the Rg Veda are based on an astronomy conducted in the Arctic and that the Vedas were point of origin for many later Indo-European thoughts about gods in the heavens. I have argued that world history is likely to have emerged from a world organized around women, with developed spiritual ideas about the goddess, the sky, and the Earth.

Any matriarchal group following the warm Gulf Stream coast would have discovered a melted Arctic Sea. The Arctic provided an extreme observatory for understanding numerical time periods as Earth rotations, lunar months, and solar years relative to other celestial periods through the long dark night of winter. This would explain why the megalithic occurred, at different latitudes, after the ice ended: the scene had been set for a matriarchal astronomy. The later flood of patriarchal Neolithic farming skills, from the Middle East, slowly destroyed the matriarchal networks of meaning, though sacred geometry, metrology, astronomy, and a strange idea of the music of the spheres were adopted as a basis for sacred buildings and stories about the gods.

While classicism pursued sacred geometry as a subject for the education of their elites, in India these ideas were still connected to their roots within an overtly spiritual tradition having impressive religious buildings backed

180

FIGURE 9.1. Angkor Wat temple, a sacred island in a square moat (visible at the top), looking west. Photo: Mark Fischer, CCAS 2.0.

by ancient texts about the numbers and geometry to be applied in temples. Buildings were influenced by Vedic stories and a cosmology that was, through its many different sects, a consistent theory of everything coming from oneness so that when India expanded into Southeast Asia in the current era, the religious buildings were similar to those of North and South India.

ANGKOR WAT

Angkor Wat was a crowning achievement of the Khmer dynasty, founded by Jayavarman II. He had returned from the island of Java to reunite his country when a previous empire fell apart around the Khmer people. The Khmer, as in Java, conflated mountains and kingship with the Hindu gods, in a royal cult with a distinct architectural style of octagonal decorative pillars and carved lintel framing of doorways, with a mountainous temple roof above a raised, multistepped platform. The world mountain is Meru, at the center of the world, and at Angkor Wat it is the tallest of five towers in the center with the outer four placed at the cardinal directions.*

These developments became distinct in the late ninth century under King Indravarman, who started the first building works at the city of Angkor, which included a great system of irrigation. Gigantic artificial reservoirs linked by rectangular grids of canals remind us of the rectangular moat surrounding Angkor Wat, all this aimed at supplying rice paddies, which then made the Khmer empire rich.

By the twelfth century, Suryavarman II built the vast complex of Angkor Wat, whose orientation was not east but west (that is, toward the dying Sun), and whose outer raised walkway takes the pilgrim on a journey through

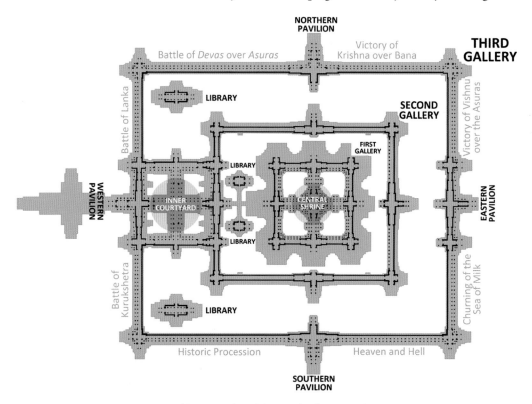

FIGURE 9.2. Locations of key mythical bas-reliefs around the perimeter include the precession of the Earth over 25,920 years, called the churning of the ocean of milk, and Vishnu versus the Asuras on the eastern, in-facing wall of the third gallery. Background plan by 3coma14 (GNU).

---

*This précis is from George Michell, *The Hindu Temple*.

eight great depictions of Hindu myth, the first of these being a historical tableau featuring Suryavarman II and past kings. George Michell reckons Angkor Wat was a "mortuary temple" akin to those in dynastic Egypt.

But this monument holds more than memorials to empire, messages for pilgrims, or fragments of world history. A deeper tradition of sacred geometry and metrology was built into Angkor Wat, encoding astronomical time on a grand, yet smaller scale than the massive yuga ages of post-Vedic literature, where the smallest of the yugas, the Kali Yuga, lasts 432,000 years. And besides, without the zeros on the end, 4320 is one-third of Yukteshwar's cyclic version of the yugas (see p. 192), once this is adapted by the traditional figure of 25,920 years for the precessional period (see the section "Water, Time, and Irrigation" below).

In our overall context, Angkor Wat concludes the story started in chapter 8 regarding the "Arctic Home of the Vedas." If Arctic astronomy sprang from Old Europe's matriarchal tribes, then, as Eleanor Mannikka points out, the founding Khmer king, Jayavarman II, married the local princess of a Shakti cult called the Nagas (or serpents), a key motif within Angkor Wat and elsewhere. Serpents are characterizations of the celestial world of cyclicity and recurrence, and one of many symbols of the goddess.

Cosmic Alignments

Angkor Wat has a strange resonance with urban building: Greek, Roman, Gothic, North and South Indian, and Jain "cities" all used accurately dressed and matching stone blocks, rather than rude megaliths. This required a high masonic culture with excellent stone-sculpting skills. But Angkor Wat also functions as a megalithic-style alignment observatory, facing the extremes of the Sun and Moon as they set on the western horizon.

Angkor Wat is an excellent observatory of alignments to the Sun and Moon setting to the west over most of the Cambodian plain. The paved walkways are ideal for accurate counting of "great time" in solar years, lunar months, and the synods of the outer planets. By contrast, Mannikkas's symbolic number lengths made for parikrama were not continuous counts, though the paved walkways were perfect for both. But counting days and months in a continuous fashion using pavements probably came first, as seen also at Teotihuacán's Road of the Dead in Mexico.* The metrology of the perimeters required additional units of measure so that, for example, the outer perimeter can then be seen as the

---

*See my *Sacred Geometry: Language of the Angels,*. 128–30.

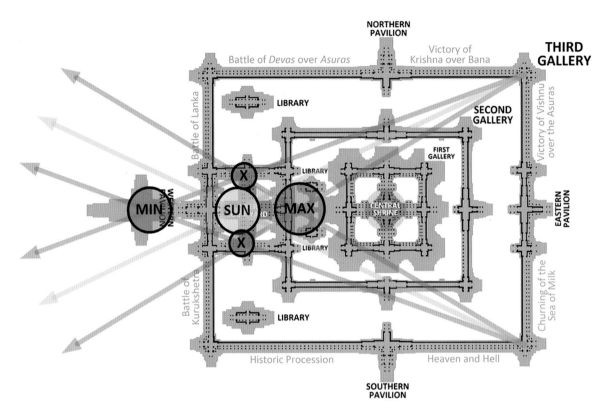

FIGURE 9.3. Image of Angkor Wat as a megalithic-style alignment observatory to the eastern horizon. Background plan by 3coma14 for GNU.

precession of the equinox in units of ten years (2592 feet of 11/12 feet) while the symbolic counts of 432 units at the bridge over the moat approaching the temple both point to the churning of the ocean bas-relief, in which counts can happily skip certain features such as thresholds and can run on into next steps or the next threshold or include a length crossing the main path, as on the bridge.

The Outer Enclosure

The metrology of astronomical counting is complementary to number symbolism, leading to traditions where numbers are related to cosmic cycles and hence considered sacred. To proceed here, one needs to make measurements of fundamental lengths within Angkor Wat, as one does with circles, squares, and rectangles within megalithic math. Guy Nafilyan spent many years surveying the monument and creating a reliable set of architectural drawings that I have used, where possible.

The temple proper has two concentric rectangles each with towers at their corners and cardinal entrance pavilions (north, west, south, and east) called the second and third enclosures. These eight punctuations of the boundaries form a crossing between two paved walkways. The perimeter pavement is crossed at right angles (under the corner towers) and the entrances to the compound crossing the perimeter walkway (at entrance pavilions). The inner walls of the outermost (third) perimeter walkway yield a perimeter of 2376 feet, which is 6 × 396 feet, numerically the mean radius, in units of 10 miles, within the now familiar equal-perimeter model of Earth and Moon.

The number 2376 feet also factors as 9 × 264 feet or 3168 inches, which is the number of the outer boundary length of an equal-perimeter model's outer circle, this seemingly defining, for the ancient world, the perimeter of the *temenos,* given traditionally to ancient sacred spaces. Therefore, there seems to be a strong relationship to the equal-perimeter model at Angkor Wat.

The Vara of 33 inches

At the same time, the perimeter length of 2376 feet divides by the number 11, being 6^3 (216) times 11. The number 11 is present in the Indian yard called the *vara,* whose standard length was 33 inches; it was the short yard of Akbar the Great and was also found in the Spanish Empire in the West. The vara is similar in length to the megalithic yard of 32.6 to 32.9 inches. A foot is 12 inches, so that a foot of 11/12 feet is 11 inches long, which, times 3, is the *vara* of 33 inches. And when 2376 is divided by feet of 11/12 feet, it gets larger at 2592 feet of 11/12 feet. The number 2592 is the head number* of the precessional period of 25,920 solar years, and this is then called a regular number, having only factors of the first three primes {2, 3, 5}. The head number 2592 is then like 396, which is in units of 10 miles, though the units of 2592 are 10 solar years. Being regular, 2592 divides by the yuga lengths, whose head numbers are multiples of 432, in this case exactly six; that is, 2592 is 6 × 432, while 2376 is 6 × 396—so that *an exact correspondence exists* between the precessional period of the Earth's axis, in time, and the size of the Earth's mean radius, in space. These relationships can be better appreciated with the help of the "matrix" diagram in figure 9.4.

---

*The use of decimals allows trailing zeros to be removed, leaving a head number and separating prime number 3, not included in base-10 numbers.

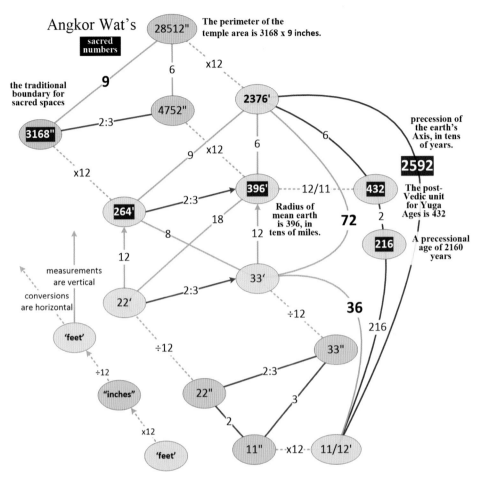

FIGURE 9.4. How the 11/12-foot subunit enabled Angkor Wat's third gallery to express both the size of the Earth as 3960 miles and it's precessional period of 25,920 solar years and the traditional yuga numbers.

Circling a Rectangle, Not a Square

The outer perimeter of the temple is a 6-by-5 rectangle that has a strange relationship to a circle via π as 22/7 since 6 + 5 + 6 + 5 = 22 in perimeter.

The idea of a *rectangle* that has the same perimeter as a circle is not very familiar. However, an example does exist near Carnac, in the form of a 3-by-4 rectangle whose 3-side is close to π at 3.142857. The width of the rectangle, being 4 units wide, can be divided by 14 and multiplied by 11* to give 22/7, the simplest approximation to π. This multiplied by 4 is then 88/7.

---

*As with the normal equal-perimeter model for square and circle.

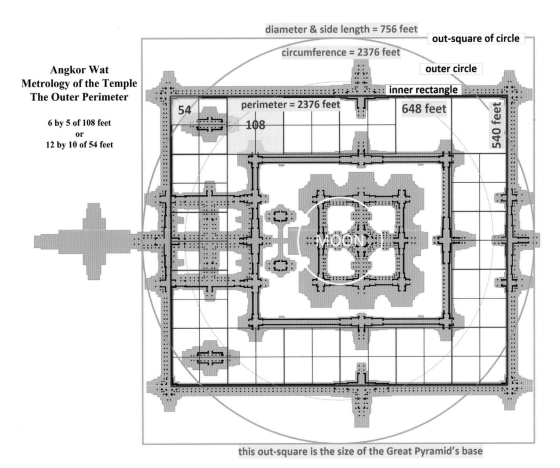

FIGURE 9.5. The inner perimeter of the outer temple boundary (gallery three), which, like the middle boundary/gallery, provided an exact 6-by-5 rectangle whose perimeter is 22: (6 + 5) × 2. It is helpful and perhaps provoking that the square temple fits within a square whose side equals the southern "index" side of the Great Pyramid of 756 feet.

A simpler way to put this is that a 3-by-4 rectangle is the perfectly fitting *home* for an equal-perimeter square and circle when the 3 is elongated to being 22/7. And at Crucuno, the stones of the southern kerb appear to have been set to both 30 and 31.42857 megalithic yards by exploiting the thickness of some southern uprights.

The situation with a 6-by-5 rectangle is different again because the perimeter of the rectangle is 22 units, and this must be equal, in perimeter, to a concentric circle of diameter 7. The solution can be visualized using the implied grid of 6 by 5 or 12 by 10; see figure 9.7 on page 189.

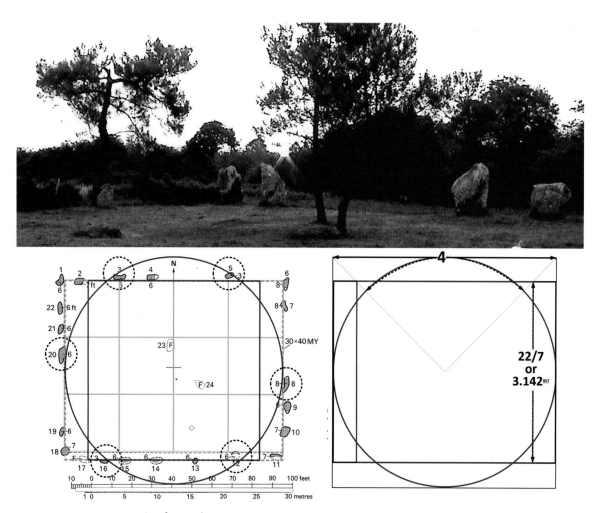

FIGURE 9.6. *Above:* the unique Crucuno rectangle near Carnac, with the Sun rising on midsummer solstice (*below left*) between stones 18 and 19. Site plan by Alexander Thom. *Below right:* the full "logic" of John Neal's proposed solution, noting that the rectangle's height "is" 22/7 relative to the 4 of the out-square.

We have therefore discovered a key to the two rectangular walkways surrounding the central temple at Angkor Wat: by making them 6 by 5, there was an implied but unrealized symbolism of circular and recurrent time cycles that their perimeters represented in Angkor feet, one-third of the *vara* short yard of 33 inches. And before looking at the inner rectangular walkway, there is the question as to how the yuga multiples of 432,000 years relate or not to a precessional cycle of 25,920 years, and its zodiacal ages of 2160 years, the same number as the diameter of the Moon in miles.

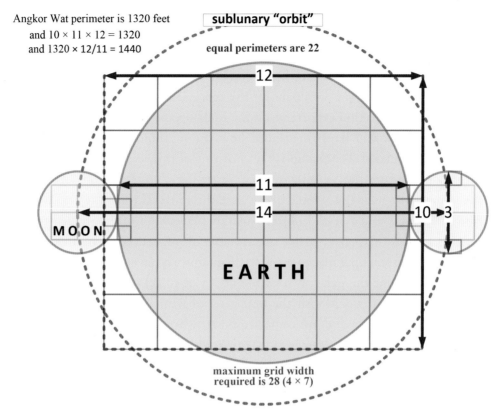

Angkor Wat perimeter is 1320 feet
and 10 × 11 × 12 = 1320
and 1320 × 12/11 = 1440

sublunary "orbit"

equal perimeters are 22

12

11

14

10→3

MOON

EARTH

maximum grid width
required is 28 (4 × 7)

FIGURE 9.7. The equal-perimeter model in the context of a 6-by-5 rectangle.

WATER, TIME, AND IRRIGATION

It is interesting why the Khmer Empire, which built Angkor Wat, arose in Cambodia. The Khmer kings had gone to Java when the dominant neighboring states and their alliances in Cambodia would have killed royal men, but the women, central to society, stayed. When the kingly heir returned from Java, there were problems providing water for rice paddies during a drought. Reservoirs and irrigation had already been developed in Java, so the returning king understood the role of reservoirs (*bayan*) and paddies. Later kings moved the capital farther south, to Angkor, and this was nearer the Tonle Sap inland sea, the fisheries, and annual floods of the Mekong, which could supply the paddies as required if reservoirs were used to store the water. As Marissa Carruthers wrote:

> In the 1950s and '60s, French archaeologist Bernard Philippe Groslier used aerial archaeology to reconstruct the layout of Angkor's ancient cities. This

revealed its vast reach and the complexity of its water management network and led Groslier to dub Angkor the "Hydraulic City."[2]

Airborne laser scanning technology (LiDAR) has shown the extent to which the empire became dominant for five centuries because of water management for rice production exportable on the Mekong and on roads.

Water use would have influenced how the mythology and its symbols could be appreciated. For example, the churning of the ocean by the gods and demons for the creation of long spiritual *reservoirs of time* and encoded with the numbers of great time would accentuate the spiritual role of water. Water flows like time, and celestial objects such as planets are often alluded to as flowing in a celestial river, in one of a number of important great circles: the rotation of the stars parallel to a celestial equator, the river of the Sun's path (the ecliptic), and the galactic river (or ocean of milk). In the precessional cycle, the meridian would visit the four gates of winter solstice, spring equinox, summer solstice, and autumnal equinox, where these rivers meet. The milky ocean in particular moves slowly in angle as the Earth's axis performs its loop over 25,920 years.

Eleanor Mannikka's deduced cubit of 10/7 feet can, through measurement, give traditional numerical meanings to the act of walking around the whole temple. Perhaps the most important fact about Angkor Wat is that the visitor entering from the west approaches the temple as a series of growing lengths that represent the yugas in reverse order; that is, backward in time, as shown in figure 9.8.

The Satya Yuga, or Golden Age (now deep in the past) is the meaning of the temple complex with its three galleries. Pilgrims travel back in time to the Golden Age, but returning travel forward through the yugas, from the Golden Age of Satya.

Indian Astronomy

By the time of Angkor Wat, Indian astronomy had been written down. It is thought that Hellenic astronomy had simply been adopted in India, but there were extra techniques for calculating time and the observance of cycles not found (even today) in Hellenic astronomy, such as the Brihaspati year of 361 days, which lies at the heart of this temple (for this, see the "Inner Temple and Enclosure" section below). Brihaspati is Jupiter, and, in an average of 361 days, Jupiter sweeps through one-twelfth of the zodiac; that is, one

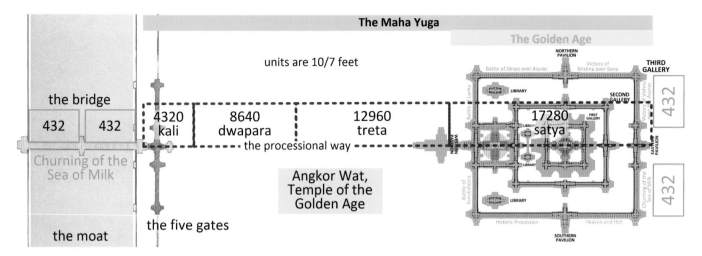

FIGURE 9.8. The central processional way first introduces the churning of the oceans with the bridge passing over the moat, and bas-reliefs there showed the tug-of-war between gods and demons, in deep bas-relief on the sides of the bridge. The remaining width of the monument, from the bridge to the end of the eastern pavilion, is then the Maha Yuga of all the yugas (1728). The units were 10/7-foot cubits of 20/21 feet.

sign of the zodiac's twelve equal houses. It was probably known in Minoan Crete but is now a unique survival within Indian astronomy. The Brihaspati year gives a concrete reason for the twelvefold zodiac as under the rulership of Jupiter rather than the Sun, Zeus having a pantheon of twelve gods. We saw in chapter 7 that some Cappadocian Christians knew of this planetary period's relationship to the Venus synod of 384 days, since the ratio between them is 55/34—two adjacent Fibonacci numbers—between our brightest planets.

The earliest description of the post-Vedic system of yugas was a set of ages, decreasing in length that, at a later time, began all over again. These yugas were initiated when Vishnu woke up and created Brahma out of his belly button, on a lotus. The yugas go through days and nights "of the gods" with themselves waking up in a golden age that ends with a dark age. These days lead to years, as for us, until Vishnu dissolves the creation and goes back to sleep.

In Greece, Hesiod's idea of world ages was also a linear one, in his day called (again) the golden, silver, bronze, heroic, and iron ages. As de Santillana points out, there were two versions of great time, one linear that restarts to the same sequence and a cyclic version that was more clearly associated with the changes in the sky over 25,920 years.

At the Arctic, some equatorial stars literally disappear, so that precession was a genuinely felt loss, yet away from the Arctic, stars sink below a celestial equator that is now elevated above the horizon. So, while old stars "disappear" and new ones "appear" at the pole, traditions farther south retain this fact when the losses and gains seem to have lost their meaning. The purported significance of the perishing and resurrection of stars, as a star cult of Great Time, seems to be an artifact of an Arctic astronomy preserved in the Vedic tradition.

Angkor Wat was laid out as a single Maha Yuga "day" of 4,320,000 years, yet the third gallery of Angkor Wat encoded the cyclic precessional period of 25,920 years. The period of precession had been adopted from the motion of the equinoctial point among the stars in approximately a human lifetime. Within any epoch, precession is the moving of the stars by a single degree* at the equinoctial point of spring. The time required for this shift was just more than the duration of a human life as "three score and ten" (70) years. Using 72 years as this marker was convenient since 360×72 years = 25,920 years.

Updating the Yugas

I propose one should forget the direct equivalence between the precession of the equinoxes and the yugas, which were *not* given a cyclic structure from golden to dark and back again. The lengths of the yugas, such as 432,000 years, would swallow the 300,000-year history of *Homo sapiens sapiens;* that is, the numbers were a code, or were "just large" like the gods, as with astronomical and geological time. We currently await the Age of Aquarius, in a *cycle* of twelve signs, but the yugas follow the scheme of lengths of 4, 3, 2, 1 units, adding up to a *linear set* of 10 units, followed by a reset.

At the Allahabad Kumbha Mela of 1894, Shri Yukteswar was advised, by the guru† of his own guru (Lahiri Mahasaya), to update the yuga system into the cyclic model with the same scale as the precession of the equinoxes, while retaining the $1 + 2 + 3 + 4 = 10$-unit structure of the yuga system. First, Yukteswar dropped two of the zeros in 432,000 years and adopted a numerically suitable (for his purposes) precessional period of 24,000 solar years, which enabled the following:

---

*Or 1/360 (1 degree) of the ecliptic.
†Babaji; see Yukteswar, *The Holy Science.*

1. The 24 in 24,000 years enabled him to integrate 12 zodiac-signed ages, that were then regular numbers.
2. The figure also allowed him to give his yugas in numbers of 4, 3, 2, 1 blocks of 1200 years each, there now being 20 units of 1200 years in the cycle.
3. Descending and ascending halves of his 24,000-year cycle he called an "electric couple," with each lasting 12,000 years. A Kali Yuga lasted one (1,200), Dwapara* two (2400), Treta three (3600), and Satya four (4800) years.
4 This meant 432,000 years (the unit for traditional yugas) had been reduced by 360.

Though Sri Yukteswar's assumed length for precession was somewhat short at 24,000 years, if one adopts his *formulation* but makes it fit the traditional period of 25,920 years, "the books then balance" quite well, as seen in the table below and in my revision of his own diagram in figure 9.9 on page 194.

TABLE OF THE GREEK VERSUS YUKTESWAR'S YUGAS, WITH CUMULATIVE DIFFERENCES

Whole period	12	6	25,920 years	24,000 years	−1920		
Hesiod	Indian*	units of 1296	signs	In descending	In ascending	Yukteswar*	Diff.
Golden	Satya	8	4	5184	5184	4800	−960
Silver	Treta	4	3	3888	3888	3600	−288
Bronze	Dwapara	2	2	2592	2592	2400	−192
Iron	Kali	1	1	1296	1296	1200	−96

It must surely be that the builders of Angkor had also obtained the 25,920-year figure and had an astronomy where the yugas seemed to point to the precessional cycle, but not for the outer meaning, which had to stick to the yugas as scripture had them. So, the architects placed this head figure 2592 of precession in years around the third gallery, in units of 11/12 feet, making the perimeter 2376 feet. In the temple, its west-to-east length conformed to the

---

*This is Yukteswar's spelling of the yuga more commonly spelled Dvapara.

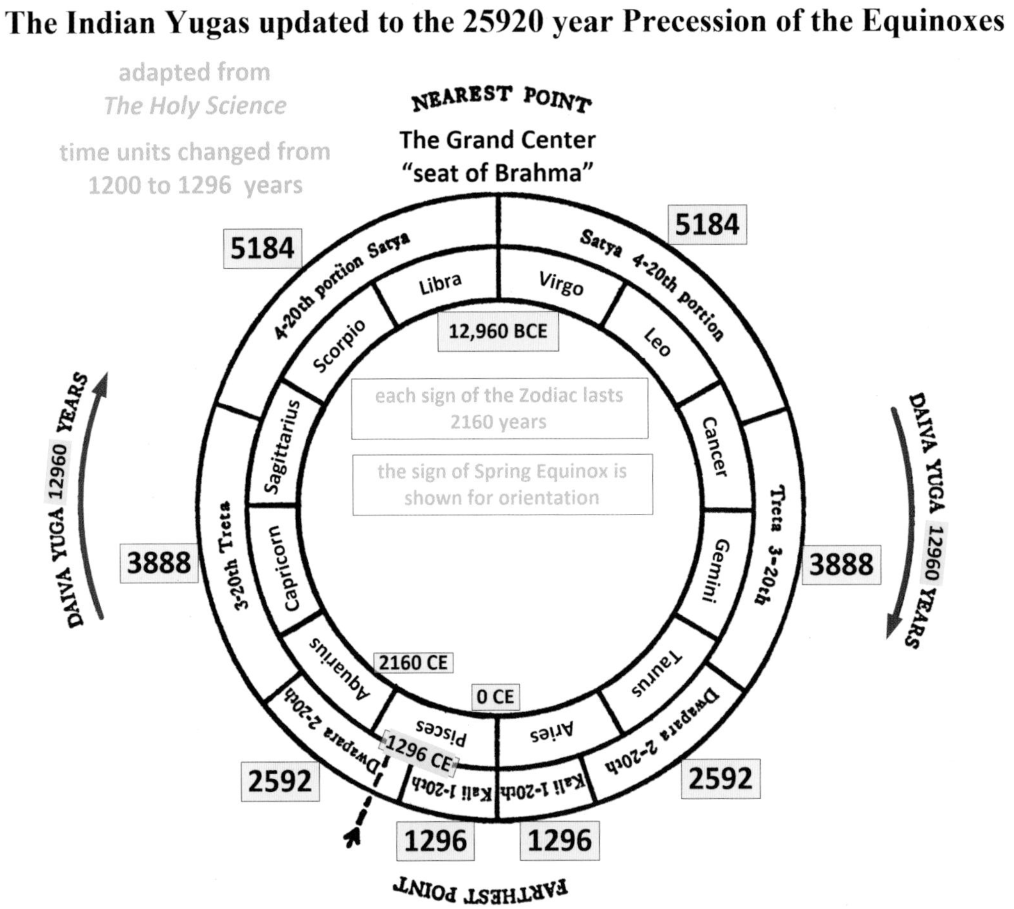

The Indian Yugas updated to the 25920 year Precession of the Equinoxes

FIGURE 9.9. Sri Yukteswar's Diagram of the Yugas, itself brought up to date by using the traditional period of 25,920 years for a complete precessional cycle. The dotted line near seven o'clock was where he placed the present position in the cycle when he wrote the book.

Maha Yuga (as shown in fig. 9.8), made up of multiples of 432, losing all three zeros but retaining their head numbers, in cubits.*

The Treta Yuga of 3888 years (of three 1296-year periods) can be seen as the god of Dwapara and Kali. Its descending and ascending Treta Yugas of 3888 years sum to 7776, the head number of YHWH (6^5 or 777,600,000)[3] and, as just the head number, Apollo in Plato's Ion: "Ion is in Apollo," since Plato's

---

*432 cubits (of 10/7 feet), similar in length to the Greco-Roman stadia of 600 or 625 feet, being 617 feet long.

Ion contains 7776 syllables.[4] This is a harmonic allusion* within the precessional framework that might suggest that the 3888 of the Treta is the "ballast" of string length, which spoke through the Bible to the worlds of the Dwapara (silver) and Kali (iron) Yugas, then leading to a date of around 4000 years BCE, for the genesis of that worldview.

THE MIDDLE ENCLOSURE

This gallery is also a 6-by-5 rectangle, which means the previous geometrical and metrological lessons apply, such as that the perimeter is 22 and a circle with diameter 7 will equal that perimeter. The inside perimeter of the gallery (as before) gives a reading of 360 feet for the northern side length, and a total length of 2 furlongs (2 × 660 feet or ¼ mile). Measured using the Angkor foot of 11/12 feet (11 inches), the perimeter is worth 1,440 of these feet (we will later see why, in fig. 9.12.)

Counting the Moon and Planets

The number 1440 is very significant in my work; for example, regarding Adam in the Bible's first chapter and the lunar counting of the "harmony of the spheres"[5] of planets relative to the lunar month.

Adam's letters have the Hebrew number equivalents of 1, 4, 40, which add up to 45, yet when spoken as "one-four-forty" they are 1440, a number that is 32×45; 32 is 25, and so there would be five octaves (45 : 90 : 180 : 360 : 720 : 1440) in between. The top two octaves—namely, 360 : 720 and 720 : 1440—are virtually identical in being able to express five modal scales around Adam. Adam was therefore sandwiched by the biblical authors (educated Jews exiled in Babylon) between the Moon (960) and Jupiter (1080), whose numbers are 9/8 apart, a musical whole tone. The cornerstone is Saturn (1024), expressing only powers of 2 (2^{10}) as Adam's tritone ($64/45 = \sqrt{2}$) relative to Adam as *do* (for the octave of 9 to 18 lunar months). The lunar years relative to Saturn synod's is then a semitone (of 16/15).[†]

In figure 9.10, I begin solving the puzzle of how this gallery could have tracked solar and lunar alignments to also see where and when full Moons occur inside the loops of Jupiter and Saturn: (1) for Jupiter, every other synodic

---

*A term used to define when harmonic numbers are found within a mythical narrative as part of its implicit meaning.

†All in just intonation, whose scales are all made up of the intervals 9/8, 10/9, and 16/15.

loop (in 27 lunar months); and (2) for Saturn, every fifth synodic loop (in 64 lunar months).

In the figure, I show these synods as if counted along the northern gallery's north walkway from northeast to northwest. In the 1440 octave, a lunar month is counted as 80 units (here, each worth 11/12 feet), and 1440/80 = 18 lunar months. If the whole perimeter is then 18 lunar months, perhaps ending at the southwest corner, then 9 lunar months would be completed at the north-east corner. This is part of the long count of the Olmec-Maya civilizations in Mexico, as the supplementary glyphs, added to a normal long count. This would

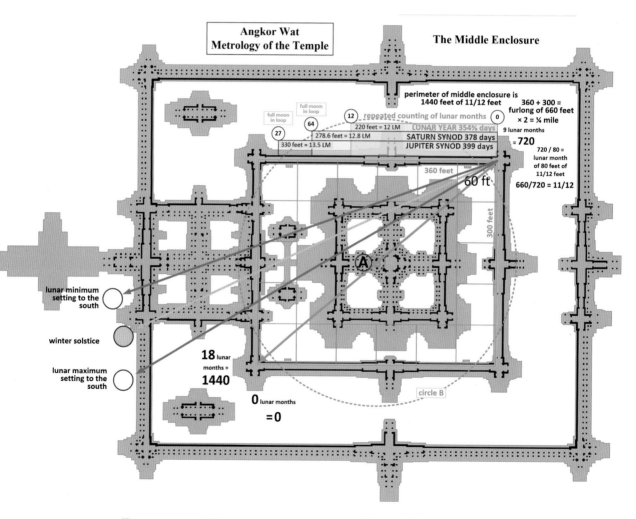

FIGURE 9.10. Harmonic interpretation of the middle enclosure. This could be something like the solar year instead for the circle and rectangle enlarged to 1461 units.

provide the location of a given moment within this additional lunar framework. And since long counts often referred to moments of an eclipse, of the Sun in particular, others have pondered whether Mexico obtained its building styles from the Indian tradition of temple designs.

However, the 18-month calendar and octave of limit 1440, where the month is 80 units, points toward a continuous counting of lunar months along the roadway that ran on indefinitely around the gallery's walkway. The punctuation of such a count with synodic loops containing the full Moon would arrive at an accurate value for Saturn of 64/5 = 12.8 lunar months and for Jupiter of 27/2 = 13.5 lunar months. Having created an instrument for the synodic harmony of the two (visible) outer planets using a simple and continuous lunar counting, the gallery was wide enough to accommodate a running count of the solar year of 365.25 days.

Simulation of the Sun's Progress

Counting the solar year is important to the life of an actual horizon observatory, and when the count is cyclic upon a circle, rectangle, or square, this enables the Sun to be tracked each year as if the walkway *were* the Sun's path, the ecliptic. We know that after 4 years there is always a leap day, caused by the extra quarter day in the actual orbit of the Earth around the Sun of 365.25 days. As we do today, a 365-day calendar can be used for three years, and the fourth can then have an extra whole day (366 days), and this sort of 4-year cycle was adopted in the classical Greek period, between the games of ancient Olympus, and two such leap years fitted the emerging 8-year calendar of 99 lunar months.*

A perimeter of 1461 units (4 years of days) allows a Sun-count marker to move 4 units every day so that each circuit was 365.25 days. This was neat, because the Sun on the horizon, on the same day each year, is staggered in just that fashion; it is 4 years before the Sun rises (or, at Angkor Wat, sets) on the same point on the horizon. And as stated above, the exact point of the Sun marker on the gallery would be the exact point of the Sun on the ecliptic, both changing together. If one could move the marker 4 times in a day in a rectangle of 1461 units to show where the Sun is on the ecliptic, then when an eclipse occurs you can know where the lunar node is.

---

*The 8-year calendar provided a seasonal calendar good for farming but in and out of sync with the phases of the Moon. At least it did not wander year to year as the lunar calendars did, for 12 months did not divide into the 12.368 lunar months of the Sun's year.

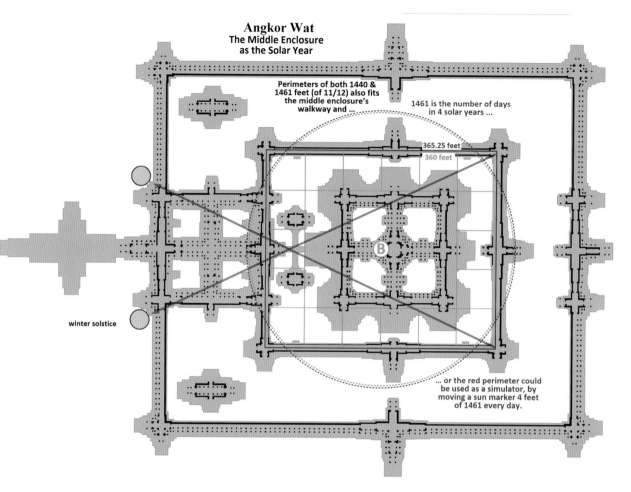

FIGURE 9.11. Using the middle perimeter walkway as a simulator for the 365.25-day solar year.

The gallery's walkway was easily able to contain the counts for both 1440-foot and 1461-foot perimeters (these counting in units of 11/12 feet). The larger count of 1461 is just 21 units greater than 1440, and the walkway could easily accommodate that.*

The 6-side of the 1461 rectangle is then 365.25 ordinary feet, which is one-quarter of 1461, just as the 6-side of the 1440 rectangle is 360 ordinary feet. The previous 360 feet had alerted me to the 1440 × 11/12 foot perimeter, the same

---

*Through an enlargement by 21/8 = 2.625 feet of 11/12 feet (2.41 ordinary feet), across the walkway.

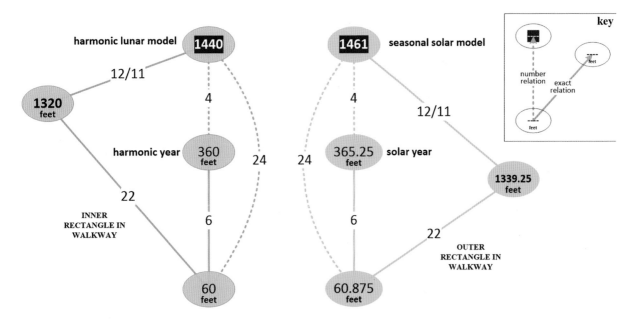

FIGURE 9.12. Angkor Wat second enclosure, matrix diagram of its two perimeters. Matrix of the middle enclosure walkway, noting the symmetry between the 6-side length, in feet, of a 6-by-5 rectangle and the perimeter of the rectangle in feet 11/12 less.

foot that converts 396 feet to $432 \times 11/12$ feet in the outer enclosure. One can see how this uncanny numerical correspondence is achieved, within measurements using these two kinds of foot, by drawing another matrix diagram (fig. 9.12).

The equal perimeter between circle and 6-by-5 rectangle enables the rectangle to be used instead of a circle. When traversing the walkway, as the Sun's path, one can use the 11/12-foot subunits (inches) to move $44/12$ ($3.\overline{6}$) feet each day, in the pilgrim's *parikrama* direction of counterclockwise.* It is therefore true that the lunar harmonic matrix of 1440 units (80 to a lunar month) and the solar Olympic period of 1461 days, could be used to count the Sun and Moon in a very integrated living calendar within the middle gallery's walkway.

---

*Parikrama is a method of prayer through circumambulation practiced in Hinduism, among other religions. It can also refer to the path itself. Many times it is practiced along a *prakaram*—an outer part of a Hindu temple along which pilgrims process in a counterclockwise direction as they approach the inner part of the temple. The movement of the planets appears counterclockwise in the Northern Hemisphere, and so it appears the Vedic pilgrim circling around the temple's perimeter was an image of time and celestial motion.

The Simulation Revolution

The advantage of counting around the ecliptic is that the moving parts are then clearly seen from the whole walkway. The ecliptic is populated by stars on it, above it, or below it. The four gates of the year, the solstices and equinoxes, the twenty-eight nakshatras, the zodiac, all could be marked so that where the Sun was could be correlated with the facts in the sky.

The location of the Sun around this enclosure, being its location on the ecliptic, would correspond to the marker stars of the Indian system of 28 lunar mansions (*nakshatras*) and to the signs of our own zodiac, these inherited from Mesopotamians via the ancient Greeks. Thus, the walkway *is* the solar year, whose first pilgrims were the astronomers keeping time with a continuous count using Sun and Moon markers. There would probably have been two node markers at opposite places on the walkway, moved when an eclipse occurs in a new location on the ecliptic. One senses compatibility with the long count of the Maya.

However, if the many eclipse markers were left in place, showing where eclipses had occurred on the ecliptic walkway, a pattern would emerge where, for example, in nearly 47 lunar months an eclipse occurs again. The best such pattern is the Saros period of 223 lunar months, after 18 years and about 10 days and after a further 12 lunar months (a lunar year), the 19-year Metonic period would have markers exactly where they had been 19 years before. Who knows what further insights would emerge from utilizing such a concrete observatory?

In the absence of our modern methods, it is therefore true that the "goddess" methods for doing astronomy were forced to use such direct methods for recording events (like event markers upon a perimeter), turning their difficulty into their advantage in *not* having to calculate everything but rather experiencing time from within a concrete simulation of the sky.

THE INNER TEMPLE AND ENCLOSURE

At last we reach the central temple, with four corner towers plus one central tower. This central complex is offset to the east within the two outer 6-by-5 rectangles. The central mass is a square laid out according to a major mandala of Indian temple design, that of Brahma the Creator God. His square has a side length of 8 squares so that 64 squares lie within the five peaks of Meru, and the perimeter of his square is half that at $4 \times 8 = 32$.

"The north Indian manduka-mandala, containing 8 × 8 *pādas*. Around the central Brahma-sthana (4 units) are the *pādas* of the 'inner gods'(2 units) and the rings of the 'outer gods' (1 unit). The points of intersection must not be interrupted by the lines of the ground-plan."

The Rules for Building a Temple of Brahma

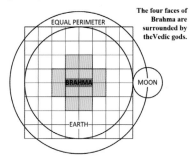

The four faces of **Brahma** are surrounded by theVedic gods.

This makes Brahma is a unique number 8, the cube of 2, but whose area (64) equals half his perimeter (32), when "pressed into the Earth" as a temple.

The Evolution of Brahma by **doubling in Area** from One

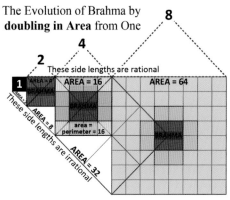

Angkor Wat as Temple of Brahma

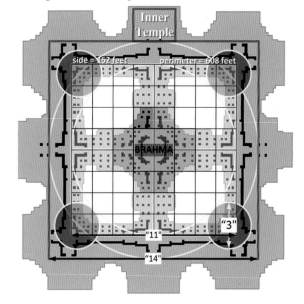

FIGURE 9.13. The inner temple as the five peaks of Mount Meru; the square (with side 8) of Brahma and his four faces; and the equal-perimeter model of Earth and Moon, with the Moon the four outer towers.

And the absolute size of this temple square is 152 feet, which is 8 × 19 feet. The area of the temple is therefore the square of this, which is 64 × 361 feet, while the perimeter of the temple is 32 × 19. The 32/64 identity of perimeter to area within Brahma's square (by making the units 19) has made the areas equal to Brihaspati's year of 361 days, still heeded by Indian astronomy. This year is even heeded by the timing of the Khumba Mela festivals, which are a religious equivalent to the Olympic games of ancient Greece, and where Sri Yukteswar received his instruction to adapt the yuga system to the precession of the equinoxes through the twelve signs, each of which is traversed by Jupiter in 361 days. While Jupiter is instrumental, the number 64 is the number of lunar months within 5 × 378 days, or 5 synodic periods of Saturn. Half this is 32 lunar months and 2.5 synods, 2.5 × 378 = 945 days, giving us an ideal lunar month of 945/32 = 29.53125 days. The design of the lunar system was Saturn who "passed the measure" to Jupiter.

10

ROME: THE GODDESS

WITHIN THE VATICAN

The square and circle appear crucial to what geometry is. Powers can be seen to spring forth from their interaction and role in modeling planetary design. For instance, the equal-perimeter model has been a haunting presence within ancient monuments. This does not sit well alongside our modern knowledge of the world, arrived at through a science based on physics and chemistry. It is okay that some very small numbers {3, 11, 14} are a geometrical approximation to π (22/7), but, when multiplied by the harmonic number 720, these numbers give us an accurate approximation to the size of the Earth—namely, its radius of 3920 miles—and the radius of the Moon, 1080 miles. It is as if this geometry had been used as a starting point for designing the Earth and the Moon. It could even be a cosmic dimensionality enabling the Earth to be the life-bearing planet it is.

REBUILDING SAINT PETER'S BASILICA

In Malcolm Stewart's book on sacred geometry, his starcut diagram was applied to Raphael's painting *The School of Athens* to reveal the radiating lines, like those of perspective, that invisibly connect the people within the picture who are standing around the Athenian Lyceum, forming a geometrical scheme. He says, "If the starcut was the central geometrical determinant for Raphael's formal depiction of classical philosophy [it was a] known authoritative device" or framework for geometrical understanding in the Renaissance. Stewart found a potential antecedent for something like his diagram in Donato Bramante's plan for a new Saint Peter's (see fig. 10.1), which was to be a square building.

203

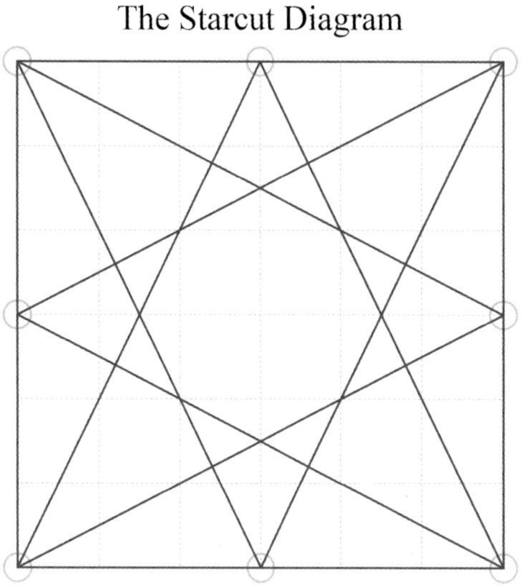

The Starcut Diagram

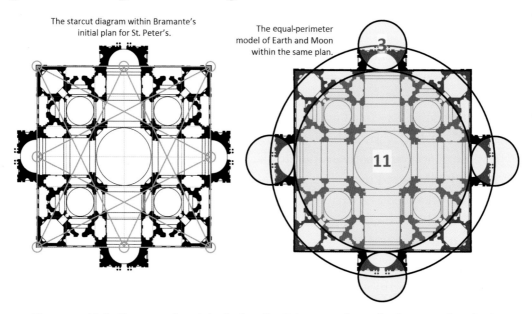

FIGURE 10.1. The simplest version of the starcut square where the sides are divided into two parts. The inner lines connect every point on the square with every other point, three points away from it. The 6-by-6 grid shown in gray demonstrates some of the many ratios available by the star-cutting of the square.

The square is again being divided into so many parts per side length, in this case two, giving four inner squares and each square can double up with its neighbors, to form double squares, and so on.

The starcut diagram within Bramante's initial plan for St. Peter's.

The equal-perimeter model of Earth and Moon within the same plan.

FIGURE 10.2. Bramante's original plan for Saint Peter's. *Left,* showing the plan's congruence with the starcut diagram but also, *right,* showing congruence with the equal-perimeter model. Both templates appear to have been influential.

Bramante's design won the papal competition, and the foundation stone was laid in 1506. However, the basilica was not completed until November 18, 1626— 120 years later—with the original plan having been extended to the south and modified under the direction of many leading artists and architects of the time.

The inference of figure 10.2 is that traditional geometry, for elite

Renaissance artist-architects, involved familiarity with a number of traditional methods, including those discussed in previous chapters. Not discussed is the starcut diagram, which can be described as a holistic interpolator using lines drawn between a square's sides, when divided by different numbers of points per side. The lines resemble an artist's use of perspective—where radiants emanate from the infinite horizon—with perspective being the key step taken in the Renaissance beyond folk art, in a search for representational realism. In the archetypal case there are three points per side, and the correlation of walkways and pillars in Bramante's design is then compelling. Stewart goes further in his figure* and finds what amounts to a design grid within the design. Making my own version from this figure over Bramante's plan, and employing modular analysis, the study of proportions, the following results are obtained. If the central cross of walkways is one-fifth of the square's side length, 5-by-5 squares (in orange) will define those walkways. But also at work is a 3-by-3 grid of squares (shown in blue) to define the central space in the standard style of the basilica from the Orthodox (Eastern Church) tradition. This would then account for most of Stewart's dotted lines.

Unfortunately, the plan has no scale from which an intended metrology

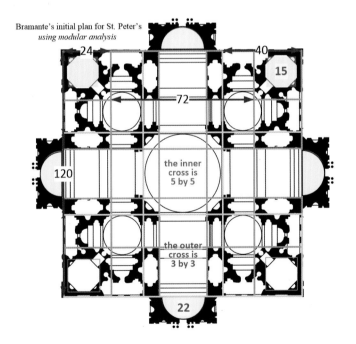

FIGURE 10.3. Analysis of Bramante's plan using multiple squares.

---

*Figure 8.18 in his *Sacred Geometry of the Starcut Diagram,* 106.

could then be ascertained. The application of units of length to problems of measurement, design, comparison, or calculation must instead be deduced. The smallest number able to hold these two grids of 5 and 3 together requires an index* of 60. To resolve the width of the corner octagons (as 15) requires the index be doubled to 120. The squares of 24 divided by the octagon width are 24/15 = 8/5 = 1.6. One can see that the starcut diagram was probably a modular design tool, now popularly used to understand the design of cathedrals, which, of necessity, can't have been designed except as meaningful wholes.

EXTENDING THE DESIGN

But this design would go through many hands after Bramante,[1] including Raphael's, Michelangelo's, Maderno's, and Bernini's, to eventually be built in a transept cathedral design (see fig. 10.4). It is therefore interesting to see what changed and even to understand why.

A later plan from the seventeenth century was made by an unknown draftsman, perhaps to record it as built and before any later modifications. The pressure to extend the building to the south is seen, in Raphael's design, intended to turn the Greek cross of the basilica design into the distinctively Latin cross by adding a nave and narthex.[†] Michelangelo instead added a raised outdoor space surrounded by continuous steps, and these steps appear shunted into the end of a nave in the finished building. But below we find Michelangelo displaying his knowledge of the equal-perimeter model but in a new variant, and he employs an unusual gift of the double square, which can create a golden rectangle, this then covering his raised stepped area.

Another papal treasure involving this geometrical model, the Westminster Abbey pavement, was explored earlier (see p. 167). At least one Cosmati master craftsman built that pavement, completed by 1268, to the equal-perimeter design. Its mosaic is depicted in Hans Holbein's *The Ambassadors*.[2]

---

*An index number can be defined as that smallest integer capable of quantifying all the main aspects of a design. The index of 60 suggests a unit of measure equal to 1.

†"A long, narrow, enclosed porch, usually colonnaded or arcaded, crossing the entire width of a church at its entrance." *Encyclopedia Britannica*.

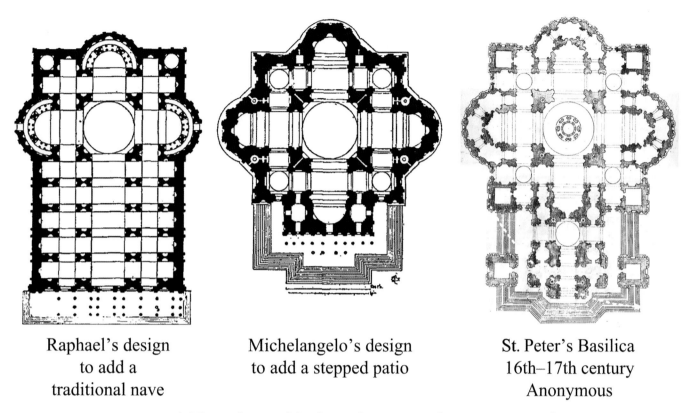

Raphael's design
to add a
traditional nave

Michelangelo's design
to add a stepped patio

St. Peter's Basilica
16th–17th century
Anonymous

FIGURE 10.4. The evolution of the design for Saint Peter's was to accommodate public space by employing two transitional designs from Raphael and Michelangelo, the latter cunningly modifying the square and extension according to sacred geometry. Right-hand plan courtesy of the Metropolitan Museum.

Michelangelo's Genius

So, while Bramante's design conforms to starcut and equal-perimeter geometrical norms, Michelangelo did something never seen before with the equal-perimeter model: he quartered the model by overlaying all four quadrants of the square so that each quadrant was a picture of one-quarter of the Earth's mean circumference. This resulted in two fish (Pisces) crossing diagonally from the corners with Saint Peter's tomb at their common center. Bramante's four-square gates were replaced with three large bays to north, east, and west, which then conform to the circular Moon of the model, but at its normal size (see fig. 10.5).

The extension of the basilica to the south seems to have called on another mystery concerning double squares within a single square, where a diagonal across the double square will, when arced to the south in this case, generate an

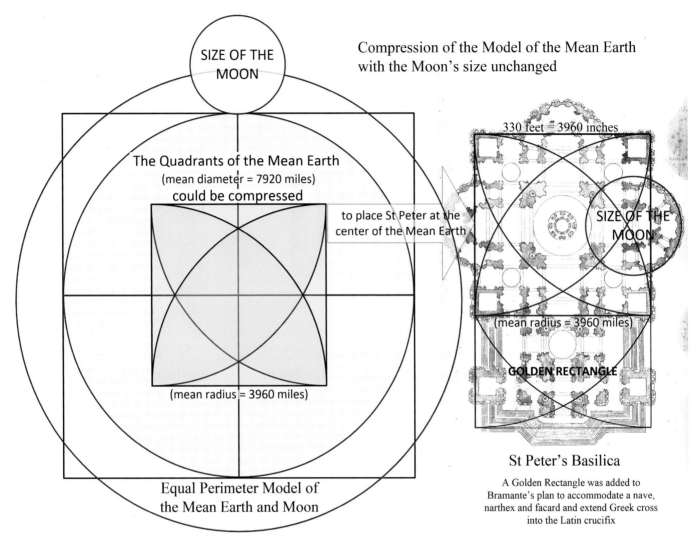

FIGURE 10.5. *Left:* the normal equal-perimeter model of Sun and Moon with four quadrants superimposed upon each other. *Right:* the retention of the true size of the Moon alongside the superimposed quadrants. (The Moon and central space were then organized as the Latin cross of a transept cathedral, giving the square its bays. The whole circumference is now within the reduced square and the Moon available for the Latin cross and bays in Michelangelo's design.)

extension, in ratio to the single square, equal to the geometrical (not Fibonacci) golden mean: a golden rectangle.

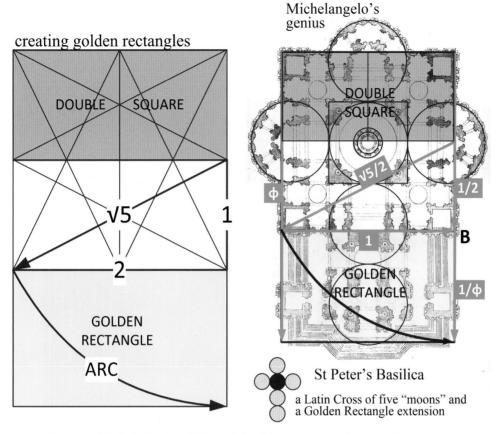

FIGURE 10.6. *Left:* the ability of double squares within single squares to create a golden ratio to the single square; *right:* how this gave Michelangelo a means to meaningfully extend the public space to the south while allowing the Moon to present an added meaning of the Latin "cross of the crucifixion."

THE VINCA AND DA VINCI

The Vinca were an extensive Neolithic people whose art has been identified as belonging to the norms of the goddess culture and who developed farming practices early on owing to their being on the Neolithic migration route from the Middle East to the Great Hungarian Plain via Greece.

Surnames came to be given when larger populations settled in towns and cities. And the name given to or adopted by people was the one most distinctive, such as where they were from, the culture or geography of their land, their trade, and so on. It is a good bet that when Leonardo's family were called da Vinci, it meant the family originated in what had been a Vinca land, perhaps in that respect unusual or distinctive. "From Vinca" would be "da Vinca," but Vinci would be the masculine plural suitable for patrilineage.

A typical Vinca figurine, like the one in figure 10.7, can be seen to betray geometric order of the triple square and, in the head, features that seem purely geometrical. This figure is wearing an apron similar to that of the Cretan snake goddess, and her arms are outstretched.

The deep resonance of the goddess, the Vinca, the Etruscans, the Minoans, and so on, had a residual presence in Italy, and this presence may be why the Renaissance occurred in Florence and Rome and not elsewhere in Europe. What would become our history was being created. And this creativity was not

FIGURE 10.7. *Left:* This statuette derives from the settlement at Vinca in modern Serbia. The flourishing Vinca culture was the largest known in Europe at that time. *Right:* The statuette conforms to the triple square as head, torso, and legs. The form of the head has a circular pate, the triangular cutaway jaw of Vinca art, and horizontal and vertical features conducive to the starcut division of the face.

based on farming and the ownership of land or possessions but instead was a freeing of latent but long-suppressed powers. Just as Indra liberated the celestial lights from bondage to the horizon by tilting the Earth's axis, Great Time creates what becomes the history out of which many wonderful things are born.

In a self-portrait as the Vitruvian Man, da Vinci speaks of equal perimeter seen in a new way, implying the golden mean extension of Saint Peter's and the raising of the circle above the square of equal perimeter within the human form. "Described by the art historian Carmen C. Bambach as 'justly ranked among the all-time iconic images of Western civilization,' the work is a unique synthesis of artistic and scientific ideals and often considered an archetypal representation of the High Renaissance."[3]

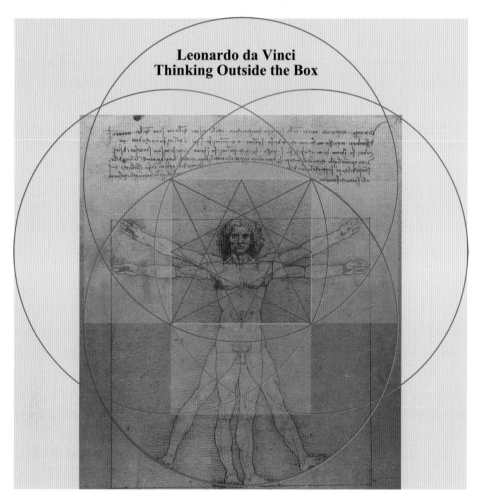

FIGURE 10.8. The strange combination of art and technical competence can be seen in da Vinci's notebooks.

11

DESIGN FOR A LIVING PLANET

The time has come to see how there must have been an esoteric tradition that understood the world as having been created by the Sky. We have seen how the Mesolithic approach to astronomy was forced to solve problems in different ways from modern astronomy, being led by a goddess culture whose workforce was not fully occupied with agriculture and so able to harvest from nature and pursue the meaning of the sky and its inherent time periods. Once their culture became synchronized to the sky and the Moon, the Moon was found to hold the numerical key to understanding, through sacred geometry, the meaning within the patterns of time. In this chapter, the Moon is found to be integrating the synodic periods of the planets through the phenomenon of the lunar month, to see how numbers and planets formed a musical scale and a ladder to heaven. This system reappeared in the harmonic cosmology presented by Gurdjieff in the twentieth century, from mysterious sources in the East.

THE CHALDEAN MODEL

According to myth, the oldest arrangement of the geocentric Earth, Moon, and seven planets was the Chaldean model—namely, the Earth, Moon, Mercury, Venus, Sun, Mars, Jupiter, Saturn, then the fixed stars. The Bible's Abraham came from Chaldea, a district including the city of Sumer, hence Sumeria, the earliest known civilization in the historical region of southern Mesopotamia. The Sumerians were a loose group of sometimes warring city-states supported by irrigated agriculture.

It is useful to see the Chaldean model according to the two types of harmony that I have found,* existing between the Earth-Moon system and the planets (see figure 9.2).

---

*During the past twenty years and as recorded in my books on different aspects of their story.

1. The **inner planets,** orbiting the Sun, are governed by Fibonacci resonances with the Earth, primarily with the inner planet Venus, whose synod is related to the practical year of 365 days, as 8 units to 5 units, the unit being 73 days.

2. The **giant planets** (Jupiter, Saturn) have synods that are musically resonant to the lunar year, respectively 9/8 and 16/15—the musical tone and semitone.

3. The **planets straddling Earth**—namely, Venus and Mars—when seen from the Earth, have synods in the musical ratio 3 to 4, the musical fourth (4/3).

These forms of harmony go largely unnoticed in the public account of modern astronomy, perhaps because numbers are not considered causative as physical laws are. Yet numbers can be existentially causative as the measurements of the synodic time periods, if gravitational resonance takes place between concentric orbits to the Sun and suborbits to the Earth of the Moon, all then seen geocentrically by us, *from the Earth:* the third planet from the Sun.

This Chaldean order of the planetary system has become associated with the astrological cult of the later Babylonians, in which the planetary aspects were thought causative in human life, but this has not been how this model survived to reach the Classical and Renaissance periods around the Mediterranean. The Bible's Jacob saw a ladder reaching into heaven with the angels traveling up and down it, now called Jacob's Ladder. This idea of the ladder to heaven, by the time of Dante, became a picture of heaven and hell upon which pilgrims and sinners were engaged in a game of "snakes and ladders."

What is extraordinary about the Chaldean model is its adoption as a picture of human transformation into the spiritual world where "the way up is the way down," so that the planetary world came to be viewed as a direct presentation of what is a religious idea connecting us to the higher worlds from which our world has arisen. Having studied some of the ideas of J. G. Bennett, a student of G. I. Gurdjieff, the world we live in can only be an act of Will to create it, but human beings are the turning point for this creation in which the creation becomes internalized as our own will becomes purified and ultimately free of the creation, which currently supports us, free to be creative ourselves and so as to add another center of creativity, in harmony with the original Will. In a sense that is the significance of the Christ as becoming the son of God through his passion, which was eventually presented in the Gospels as his story on Earth.

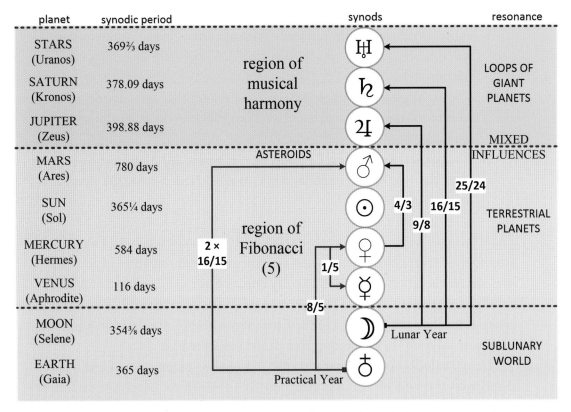

FIGURE 11.1. Geocentric entities, synodic periods, and the two types of harmony. The planet Uranus is hardly discoverable by naked-eye astronomy, yet it, too, has a key musical interval relative to the lunar year. See 11.2 for a more holistic version.

In Bennett's well-formed language the creation is the result of the Universal Will while the purpose of the creation is the formation of the Cosmic Individuality. Bennett describes individuality as a manifestation of harmony, through which the facts are harmonized with the values implicit within the creation. The Sun and planetary system are then a third type of Complete Individuality, while human beings have a potential individuality called the True Self—a work-in-progress within a phenomenal planetary world, which faces our selfhood. Figure 11.2 presents a simplified view of the planets connecting the lower world of life, imaging the higher worlds, and the potential life of the transformed human who has grown a spiritual body through a developmental process often presented as rising up through planetary levels, as seen in alchemy and the Mithraic cult that vied with early Christianity as a new religion for the

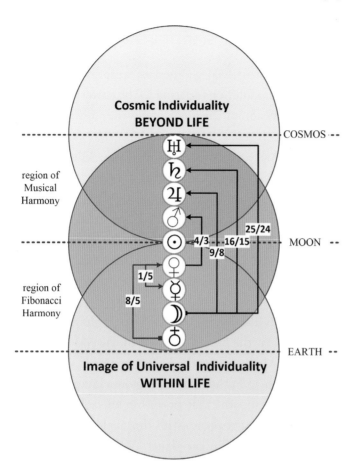

FIGURE 11.2.
Diagram of Universal
Individuality within
Nature, and Cosmic
Individuality outside
Nature, bridged by the
Chaldean planetary
model, with Fibonacci
and Musical harmony
in either vesica of
overlapping influences.

elite. By the Middle Ages, the Roman Church ruled, but the Chaldean model and Pythagoreanism lurked among the educated classes, hence Dante and many later poets wrote within the misty gloom of ethereal realms, conjunct with salvation and damnation.

We cannot here describe how the Chaldean order is thought to have arisen, perhaps alongside the seven-day week of planetary naming, but perhaps the faster moving planets were assumed nearer to the Earth. But it is clear why there were two regions of harmony, with the terrestrial planets in the first part followed by the slower outer planets. The Fibonacci numbers between the Earth year (365) and Venus synod (584) are due to gravitational resonances as a result of their proximity. But the musical harmony to the outer planets is not due to their orbits alone but instead to how we see the other planets from our moving planet and how this harmony is then expressed relative to the lunar month and year, rather than the solar year.

THE MOON AS FOOTBALL FOR THE GODS

I have previously shown that two-dimensional geometries were used, in the Stone Age onward, to provide numerical solutions relating to the patterns within celestial time—for example, the precession of the lunar nodes over 18.618 years relative to the solar hero period of 33 years. The relationship in that case is the equal-area model of a circle radius 18.618 units to a square of side 33 units. So 33 years divided by 18.618 = 1.7724 . . . which squared is 3.1416 . . . that is, π more accurately than my calculator can achieve it, it is exactly π!

A similarly simple relation exists in the simple numbers of lunar months when seen as a harmonic model, linking the Earth environment to the planetary worlds through the Moon. In this, the cornerstone is the lunar month, which equals 1 for that model based on whole numbers and their ratios as musical intervals. The lunar month is the synodic period of the Sun to the Moon, in her orbit, seen from the Earth through illumination of the Moon's phases by the Sun. For example, the full Moon, when the Sun is opposed to the Moon, with the Earth in between and; the new Moon, when the Sun is conjunct with the Moon, with the Sun behind the Moon and eclipsed if sitting on a lunar node.

The lunar year is the twelve *whole* months that fit within the solar year of 12.368 months, and it is this lunar year that resonates with the giant outer planets (Jupiter, Saturn, Uranus) and with our "sister" inner planet (Venus). As we shall see, a musical scale is created by the lunar years that we would call the major diatonic (the Ionian scale). This scale has special properties not available to the ancient world's heptatonic scale (preferred by Pythagoras), and today it is called the Dorian scale. Quite unlike our major scale, the Dorian is symmetrical around its tonic "do."

It is the major scale that enabled our modern music making, achieved using keyboards, staves, and equal temperament, where each key of the medieval keyboard became able to transpose the major scale into any of its starting "keys," this then requiring equal temperament for any key change (called a transposition) to be guaranteed harmonious. In the case of the Moon, the creation of the major scale was numerically possible only after 2 lunar years of 24 lunar months had elapsed. However, time is a continuous dimension, and so each lunar month has a month 24 months earlier and 24 months in the future.

The double synod of Jupiter then lands upon 27 lunar months (an interval of 9/8, each Jupiter synod being 27/2 = 13.5 lunar months). The next tone number is 30 lunar months (an interval of 10/9 and, within the octave 24–30, 5/4, the major third). Beyond that is a semitone interval (of 16/15) reaching 32 lunar

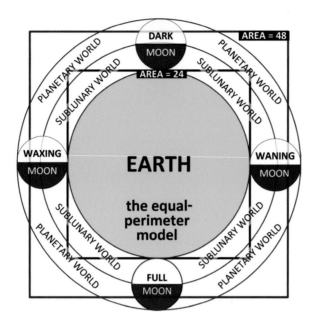

FIGURE 11.3. The Sun positioned over the lunar month. As the Sun illuminates the phases of the Moon, the Sun and Moon define the synodic periods of the planets. These synods then define harmonic intervals to the lunar years of 12, 24, 48, and 96 lunar months.

months. This is 5/2 synods of Saturn so that, in the octave 48–96 lunar months, 5 whole synods of Saturn equal 64 lunar months, making the synod itself 64/5 or 12.8 lunar months long.

To summarize so far:

1. In this raw but continuous lunar octave (24–48), the Jupiter synod resonates with the (so-called) second note (27) after do (24), the first.
2. The Saturn synod resonates with the fourth note (32).
3. The fifth note is then 3 lunar years (36) and,
4. after a further just whole tone (10/9), 40 lunar months exceeds 2 Venus synods (39.55 lunar months) by 81/80. This slight lack of pure integer harmony is due to Venus being primarily resonant with the Earth year of 365 days and less resonant with the lunar year of 354.367 days.* The exact ratio (81/80) is necessary to relate the two forms of harmony,

---

*My book *Harmonic Origins of the World* can provide the reader with a lot of further aspects of ancient tuning theory.

this being the ratio between notes tuned to the just and the equivalent tones in Pythagorean tuning.

5. The seventh note (45) is the number of Adam in the Bible, special in expressing 3 squared (9) × 5. The Vedic Prajapati* was a similar concept: his temple plan is 9 by 9 (or 81) rather than that of Brahma, the creator god (see fig. 9.13).

INTEGRATING GRAVITY USING SQUARE AREAS

Temple designs that employ grids of squares within a sacred geometry were perhaps a primitive link to the two-dimensional area functions of Kepler's planetary laws where the areas swept out by the elliptical planetary orbits around the Sun at one focus, in a given time period, are always equal. In modern differential mathematics, this could equate to the integration of forces over time, these acting on the Moon as a cross-multiple of planetary force and time, which resulted in Kepler's area law of planetary motion but here are acting continuously upon the Moon as a body free in space.

The circle is the epitome of a celestial orbit while the square is epitome of an area: the square root of a square's area is, by definition, just that, its square root. In the case of Brahma's temple grid, the area is 64 so that the side lengths of the square are 8 units, this reminding us of the 64 lunar months taken by 5 Saturn synods. In contrast, Prajapati's temple square of 81 has side lengths of 9. The circle and the square are shown related in the square and its in-circle, where the circle's diameter is the side length of the square. If that side is called 1 in length, the perimeter of the square is 4 but the circumference of the circle is π (3.1416 . . .). The ancients often used 22/7 to approximate π, making the perimeter 4 larger than the circumference by 6/7. As stated earlier, this modeled the relationship of the Earth diameter to the Moon diameter as being in the ratio 11 to 3, as seen in the equal-perimeter model.

With this in mind, the Moon is orbiting the Earth while episodically approaching the Sun (every lunar month) and opposing the Sun (during every loop of each planetary synod, as per fig. 11.3). Because of this, the common

---

*Prajapati is the Vedic sacrificial god whose very being redeems the World for its cosmic sins, sins in the world of tuning as inevitable being between the Essence (the cosmic Will) and Existence. Eighty-one is also the numerical value of Eve's name, as the "mother of all that lives," perhaps the eleven essence classes of Everything Living on the Earth and in the planetary worlds, within the supreme reality of the mother goddess.

factor of the lunar year has become numerically entwined in highly significant ways with the planets, and I believe, this led to the entire tradition that numbers and geometries were sacred. The root of the simple musical ratios (27, 32, and 40) relative to the lunar years (12, 24, 48, 96, etc.) must hail, being integers, from their being areas (as per Kepler's laws, but continuously acting on the Moon). The roots of the squares are their side lengths, which, more often than not, are irrational (only 36 being 6 squared).

This allows us to see that the number field, while itself abstract, is *far from abstract* when actualized within a musical instrument (or resonator) such as a piano or a guitar. With planetary periods, string lengths are instead their average synodic periods, and these orbital details *had to be actualized*. This is a subtle point: though the integer relationships within the number field are real, they are not actual without realizing them within existence.* The irrational side lengths make it possible for the physical world to realize integers that are themselves areas, rather than lengths, within the gravitational environment of the Moon, where many planetary forces vary over time but are being integrated over time, by the Moon! In this respect, the lunar numbers of months are achieved differently by each planet but within a common scheme. It is this that created "the music of the spheres," seen from the surface of the Earth, that was construed to form a pathway to heaven.

If the lunar month is the cornerstone of the octave (that is, the common unit of area 1) then 24 lunar months, 2 lunar years, is a synodic cycle of the Sun and Moon, which can be seen as having an *area* of 24, then capable numerically of developing the smallest possible major diatonic octave of {24 **27** 30 **32** 36 **40** 45 48} lunar months.

The sides of the area, 24 lunar months, must be the square root of 24, which is $2 \times \sqrt{2} \times \sqrt{3}$ (4.8989 . . .).

a. The area **27** has a side length of $3 \times \sqrt{3}$ (5.1961 . . .),
b. **32** has a side of $4 \times \sqrt{2}$ (5.6568 . . .) while,
c. **40** has side length $2 \times \sqrt{2} \times \sqrt{5}$ (6.3245 . . .).

Instead of the primes {2, 3, 5} that make up the octave integers, it is the roots of these early prime numbers that become the irrational factors {$\sqrt{2}$, $\sqrt{3}$, $\sqrt{5}$} of the

---

*The language here is important. It is normally assumed that abstractions are not real, yet when numbers are present as distances, the numbers affect us in an existential way. It is then that abstractions have been actualized and that the properties of number have become real. The numerical structure of time on Earth is real and has been actualized to express the abstraction of harmony in the number field on Earth.

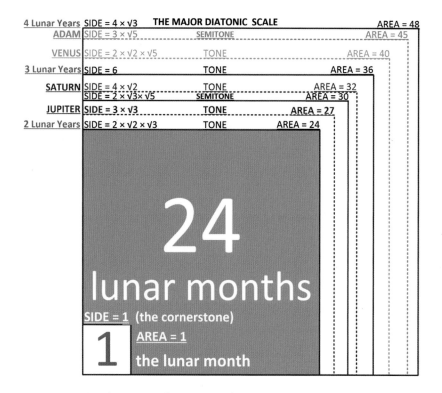

FIGURE 11.4. The major diatonic scale existing between the cumulative influences of the sun and planets upon the Moon. The units of area are lunar months and the largely irrational sides of the squares are susceptible to geometrical generation (see figure 11.5).

side lengths. The golden mean and Fibonacci integers are normally seen apart from musical harmony since the golden mean has its roots in the square root of 5 ($\sqrt{5}$), so entering into the octave {24 **27** 30 **32** 36 **40** 45 48} seen as a set of square areas.

One cannot avoid comparing this scheme to Gurdjieff's 1917 cosmology in which he introduced a set of Worlds emanating from ONE (like the lunar month), and generating six further Worlds {3 6 12 24 48 96}. Thereafter, these numbers in some way created a pattern of descending major diatonic octaves within octaves,[*] in which Life was a "side-octave" emanating from our star, the Sun, as world 12. The Moon was the root of this side-octave, in a double octave[†] between 24 and 96. The above picture is the same, in which lunar months seen

---

[*]A descending diatonic is a Phrygian scale of S T T T S T T, which in ascending is the diatonic Ionian scale.

[†]Ancient Greek music theory employed a double octave, with middle tonic *mese* or "middle."

Geometrical Generators of Irrational Sides for Rational Square Areas

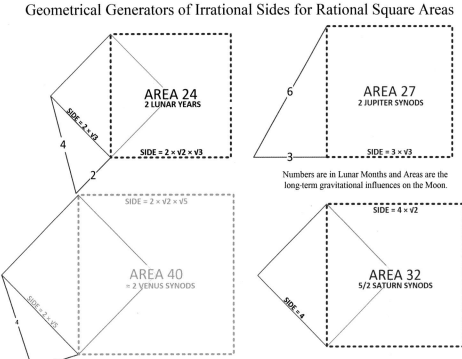

FIGURE 11.5. The geometric generators required to realize the 1st, 2nd, 4th, and 6th notes of the major scale corresponding to the double lunar year (24), 2 synods of Jupiter (27), 2.5 Saturn synods (32), and 2 Venus synods (40), in lunar months.

from Earth are part of a celestial musicality reflecting the planetary cosmos.

The origins of our own explanatory system and the early version presented by Gurdjieff must have preceded the development of the main mythical frameworks that we have found to be a virtually worldwide (human) phenomenon.* Myth and its major themes largely flow from this cosmology of the Moon, Sun, and planets. Gurdjieff asserted that "nothing happens upon the Earth by accident" and that "life on Earth is a similarity to the already created" *due to the Moon;* that is, according to laws already forged during the creation of the universe, to maintain the implicit order we see in the cosmos, as an intelligible

---

*This thesis was latent as far as the mythical content was presented in *Hamlet's Mill* but without more than the suggestion that it is the numbers, the measures, and the geometries that must somehow lie behind the myths, for there to has been commonality of themes.

phenomenon, yet conforming to the physics of it. Musicality is a form of created order that gives time a holistic structure of coincidence, which is perhaps its primary Purpose. Musical works exploit that power to make harmony, which, in Bennett's view, lies at the heart of individuality as reconciling fact and value.

THE GRAND DESIGN

But there is the question of why the Earth has the planetary environment, whose strongest symbol appears as the equal-perimeter model of Sun, Moon, and Earth. The answer appears when the diatonic square areas of figure 11.4 are made concentric around the area of 24 lunar months, which square is then the out-square of the circle that is the mean Earth (in cross section—figure 11.6). This then appears to be the composite picture of both time and space as a design for our planet.

The lunar month is the illumination of the Moon by the Sun, so that it is a synodic period unlike the Moon's orbit, which belongs within existence and the world of fact. The Moon's phases are a synthesis of the Sun's changing location in the zodiac and the Moon's orbital motion.

We saw in Angkor Wat (chapter 9) that the Earth can be seen as 396 feet in radius as well as 432 feet of 11 inches (11/12 feet), 432 being the head number of the yuga system of great time. For this reason, time and space run parallel in the equal-perimeter model as a design, typical or unique, for a living planet. An immense amount of celestial mechanics would be required to create such a Moon and Earth and then have planetary orbits (around the Sun) that are organized in a way preparatory for the expressing of harmony using the Moon as an orbital resonator, with periodicity equal to the only recently achieved lunar year of 12 lunar months, or 354.367 days; that is, the Moon behaves like a planet whose orbit of the Sun would be 354.367 days, yet no such "harmonic planet" exists. Instead, the Moon has achieved such a time period as the length of twelve lunar months (the lunar year), resonant with the planetary synods uniquely seen from the Earth. The only way to explain this is to say that the solar system is holistic in the sense that the idea of it, seen within the equal-perimeter model, created it out of the solar nebula. This narrative is incomprehensible to modern science as presently configured, being self-limited to the study of only laws and unguided forces.

It is now possible to see how the harmonic model completes the equal-perimeter model and allows us to overlay the details of the planetary octave.

Figure 11.6 is a square version of the so-called Chaldean model that lay behind spiritual literature based on the previous cosmological norms of a geocentric model of the universe; that is, the Jewish, Christian, Islamic, Buddhist, Tibetan, Hindu, Pythagorean, Platonic, yogic, shamanistic, and many other primary and secondary texts, including Shakespeare and the Western poetic tradition, alluded to the details of this geocentric cosmology: as a foundational framework now often debunked as inaccurate, yet the geocentric was a valid ordering of the planetary cosmos as a creation of higher intelligence. The religious values within the model were largely discarded and obliterated by a new scientific norm, one that was inherently materialistic in its study of physical laws and processes. This norm was thought to be universal throughout an ever-expanding vision of the universe—a vision explored using telescopes,

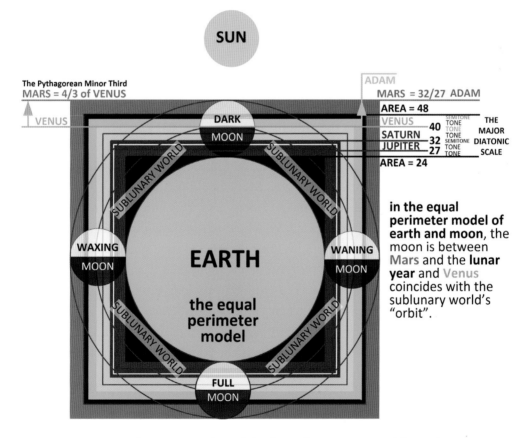

FIGURE 11.6. The integrated model of equal perimeter and lunar harmonics, looking exactly like a Mandala. Over the lunar month, the Sun illuminates the phases of the moon. In each lunar month, the area swept out captures influences form the planets, which, relative to each other, express a diatonic octave.

experiments in physics and chemistry, and deducing many new types of knowledge. The industrialized West soon used these new laws and discoveries to build new empires, exploiting the material properties and laws revealed. When the cleric Copernicus suggested everything did indeed revolve around the Sun and not the Earth (and posthumously got away with it), the geocentric past was ditched, like the baby with the bathwater, so that *the whole of spiritual literature,* in all its shapes and sizes, lost its very foundations.

The area 24 to 48 lunar months has two major regions called in the geocentric model the sublunary and planetary worlds (shown in fig. 11.6), and the diatonic octave of the planets expands from the surface of the Earth to exceed the four Moons shown as dark, waxing, full, and waning. Because Venus is nearly 40 lunar months, and the synod of Mars is 4/3 of that (shown stone colored in fig. 11.6), Mars exceeds 48 lunar months (in square area) but exactly embraces the farthest tips of the four lunar disks. Also associated with this outer extent of the equal-perimeter model is its ideal diameter of 17, pointing to the 17 synods within the 6800-day period of the lunar nodes (see fig. 8.3). In the Vedic tradition, Mars is the son of Shiva, called Kartikeya. He has six "heads," which are the six whole loops Mars describes in the zodiac, between his synodic periods (of 778.56 days), indicating how multiple heads can be astronomical identifiers. Gurdjieff placed Beelzebub, the wisdom character in his epic *Beelzebub's Tales to His Grandson,* on Mars, while Adam is 45, within the octave 24 to 48.

It is so unlikely that the octave plus Mars should define an area that forms the out-square of the equal-perimeter model, that this composite model has brought us into the realm of a completed whole, where the eight anciently visible planets are represented by the area (in lunar months) of their synods. These areas populated the numerical range of lunar months between 2 and 4 lunar years, and equally 4 to 8 lunar years (and so on) for a very long time, since the Moon became resonant with the planets when, coincidentally, *Homo sapiens* appeared.

The square area of Venus (40) is similarly interesting since she runs through the center of the four Moons; that is, she is the equal-perimeter circle, central to the lunar "orbit," while also being the young queen goddess, the Earth being the mother goddess of all life, and the Moon the wizened goddess of wisdom. If you wanted a single image to present the geocentric model of the Earth within the planetary world, this one is so good as to imply that the planets do indeed create the narrative structure found within an octave through their average individual actions, at intervals, with the lunar year and relative to each other's synodic periods.

By so easily discarding the geocentric imagery prevalent within millennia of cosmological thinking, a world of meaning collapsed, losing the relationship between fact and value. Thereafter, this model was put down to prescientific fantasies of the less-advanced past societies. This pattern of loss is clear in the fate also of the matriarchal societies that populated prehistory before the advent of the Neolithic farming that made large-scale civilization possible. We have demonstrated that Mesolithic societies left their intellectual life in their cosmic monuments and in the tools, techniques, and myths that fed into later civilizations, when they arose to replace them from the surpluses provided by agriculture.

The heliocentric and geocentric models of the planetary system were two views of the same phenomenon, but the heliocentric was taken to be the only adequate view of reality, and, because of that, the spiritual view connecting the Earth to the cosmos was rejected by the recently civilized mind. This shut down the meaning of spirituality for the vast majority, who instead serviced desires for economic growth and an improved standard of living. If religious, heaven was translated into exactly nowhere, reliant on absolutes such as God, or texts that must not be altered by those experiencing the spiritual world. This latter-day witch hunt, an extension of the feudal system, could be corrected if academia were to accept that the world has a spiritual purpose at odds with the human purpose of exploiting the planet, and indeed the planets, for their material goods.

The integrated model of fig. 11.6 allows us to see that our world was created by higher intelligence, associated with the planetary system. Everything done to break it will leave humanity poorer and then scrambling to replace those supports nature provided for our evolution. Who wants to replace the natural world with a wholly man-made world?

POSTSCRIPT

THE DEEP SIGNIFICANCE
OF OUR PAST

Not only do we wonder what ancient people were up to, but this book wonders why we cannot pass through the self-imposed glass floor of history and understand prehistoric monuments and other traces left by our prehistory. And it seems clear we have lost an aspect of ancient human beings through our successful enterprise of controlling the physical world. Formerly, knowledge was not the controller of what was thought about the world, and, in some certain way, lack of such a knowledge allowed views to develop in which the physical world was a reflection of another world. The obvious reflection of the surface of the Earth receding beneath our feet is the sky above us. And in some way our modern knowledge of the sky denies that way of thinking, which was poetic: that the structure of the world is a manifestation of the sky according to some unseeable Will. The knub of this difference of views, ancient and modern, lies in who we think we are versus the universal creativity who made us. Compared with that, our will is weak but assertive in its property rights over the Earth, and soon the planetary world.

Here we propose history has a form that has been guided by one or more groups of human beings in touch with higher intelligences, to support the evolution of Stone Age humanity and their understanding of their sacred human function, to understand the sky and Earth. We have sketched out a means whereby the sky could have been understood to have, in space and time, a sophisticated numerical structure, using an astronomy manifested within stone monuments. This work appeared to end when the predominantly matriarchal Mesolithic tribes were displaced by the Neolithic revolution of farming, which spread from the so-called Middle East to Central and then the whole of Europe. But the effects of ancient astronomy became the bedrock from which later civilizations forged their religious meanings found in the sky upon the Earth in their own sacred architecture. The new religious enterprise was then reflected upon the Earth just as the Earth was a manifestation of the sky. This marks the beginning of our history, which has slowly forgotten that the originating source of all creation had a form written in the sky and discovered before history began.

Some of the patterns written in the sky have proved both central in meaning and widespread within the historical period—for example, the equal-perimeter model of mean Earth and Moon, equal-area model of the lunar nodal period and solar hero (33 solar years), and the harmonic model of planets relative to the lunar year. The equal-perimeter model is very widespread, perhaps even inspiring the pecked cross of the pre-Columbian Mexican civilizations. These models appear to have spread as an image for a spiritual center upon the Earth, and as the last chapter works out, all the models can fit within its single framework to become an image we call a mandala, with four gates and multiple concentric squares and circular orbits. The concentric squares present how the Moon absorbs (or integrates) the gravitational tug of the inner and outer planets, whose frequency, relative to the Moon and each other, places them in a major diatonic scale of pure tones rather than equal tempered ones.

And it is from the planetary octave between Earth and sky, mediated by the Moon, that the growing individualization of human beings since classical times (reflected in many extraordinary teachers around 600 BCE) became crystalized in versions of Pythagoras's picture of man as microcosm, seen within such mandalas, systems of chakras, and narratives of the possible transformation of human beings upon a ladder between heaven and Earth. Texts* fulfilling Jacob's vision of a ladder† upon which the angels and presumably humans, having seen it, can move upward and downward. But the literal meaning of these images as imminent in the sky was, at the same time, increasingly being lost. Yet the ancient numerical and harmonic connection between the Earth and the sky still exists, involving the Moon and planets, this sweet in the mouth but bitter in the belly,‡ meaning hard to realize given our current beliefs.

---

*Using the new alphabetic writing rather than spoken words to create what it termed "The Word," in Greek *Logos*.

†"And [Jacob] dreamed, and behold a ladder set up on the earth, and the top of it reached to heaven: and behold the angels of God ascending and descending on it." Genesis 28:12.

‡"And I went unto the angel, and said unto him, Give me the little book. And he said unto me, Take it, and eat it up; and it shall make thy belly bitter, but it shall be in thy mouth sweet as honey." Revelation of St. John of Patmos 10.9.

Here, our three remarkable conclusions are:

1. That the lunar month, and year of 12 months, was created and adjusted to conform to the early number field and its early onset of musical harmony.
2. That the Earth and Moon were then conforming geometrical and harmonic possibilities expressed in prehistory using the circle and the square.
3. That all these numerical coincidences within space and time apply to human life upon the Earth and skies as shortcuts that enrich our *possibilities,* making life richer than would be the case on another planet.

One can reframe the planetary world: from being a mechanical determinator of human fate into an inherent and tireless work of creating potentials and opportunities for human life. The bare bones of cause and effect could be exceeded by gravitational objects to create a transfinite and diverse action upon the Earth, still expressing the wholeness of a single turn (the "uni-verse") of meaning-making, visually apparent in the starry worlds beyond the path of the Sun, Moon, and planets. This adventure in meaning is from the individual outward as a voyage of self-discovery that turns out to be the purpose of the universe to create human beings. This would be a project hatched by the universe itself: to create upon some planets that type of *being* whose destiny is *becoming,* and hence having a divided or incomplete selfhood.

> *For now we see through a glass, darkly; but then face to face: now I know in part; but then shall I know even as also I am known.*
> —1 Corinthians 13:12

Appendix

Astronomical Periods and Their Durations

Days and years are solar unless otherwise stated.

Period	Duration
Sidereal Day	0.997283 day or 365/366 days
Sidereal Month	27.32122 days or 82/3 days
Synodic Month	29.53059 days or 945/32 days
Lunar Year	354.367 days or 2835/8 days
Solar Year	365.2422 days or 12,053/33 days
Saturnian Year	364 days or 52 weeks
Saturn Synod	378 days or 54 weeks
Jupiter Synod	398.88 days or 57 weeks or 400 sidereal days
Uranus Synod	369.66 days or 1109/3 days
Lunar Nodal Year	6800 days
Solar Hero Period	12053 days or 33 years

NOTES

INTRODUCTION TO PART ONE

1. Paulsson, *Time and Stone.*
2. Robin Heath, (1993, 1998, 2004, 2010, 2014, & 2016), see also Wikipedia, "Stonehenge."

1. THE LANGUAGE OF THE MESOLITHIC

1. Cunliffe, *Europe between the Oceans,* 105.
2. Paulssen, *Time and Stone.*
3. Paulsson, *Time and Stone.*

2. THE MEDITERRANEAN TRADITION

1. Richard Heath, *Sacred Number and the Lords of Time*, 19–21.
2. Haklay and Gopher, "Geometry and Architectural Planning at Göbekli Tepe."
3. John Michell. 1991, 1994, 2008.
4. Haklay and Gopher, "Geometry and Architectural Planning at Göbekli Tepe."

3. REUNION OF THE MEGALITHIC LATITUDES

1. Richard Heath, *Sacred Number and the Lords of Time,* especially chapters 3 and 5.
2. Paulsson, *Time and Stone,* fig. 1.4.
3. See bibliography for Robin Heath: *Bluestone Magic,* 2012; *Proto Stonehenge in Wales,* 2014; and *Temple in the Hills,* 2016.
4. As defined in Richard Heath, *Sacred Geometry: Language of the Angels,* 237.
5. Exhaustively documented by Peter Tomkins in his *Secrets of the Great Pyramid,* 1971.

6. Described in Richard Heath, 2014, chapter 8, *Designer Planet*, 205–11.

7. See Neal, *All Done with Mirrors,* 114–19.

8. Richard Heath, *Sacred Number and the Lords of Time,* 207–11.

9. Thom and Thom, *Megalithic Remains in Britain and Brittany,* 145.

4. TIME, GENDER, AND HUMAN HISTORY

1. Ouspensky, *In Search of the Miraculous,* 56, 58.

2. Michell, *Twelve-Tribe Nations.*

3. Microbius, *Conviviorium Saturnaliiorum Septem Libri.*

5. MATRIARCHAL CRETE IN THE BRONZE AGE

1. See Hesiod's *Theogony.* The first primordial gods (116–36) follow, along with the tale of the castration of Ouranos (137–87) and its outcome, including the birth of Aphrodite (188–210).

2. Graves, *The Greek Myths,* sections 6 on castration and 7 on dethronement.

3. Banou, "Minoan 'Horns of Consecration' Revisited."

4. Pendlebury, *The Archaeology of Crete,* 100, fig. 14.

6. PYTHAGORAS, THE BIBLE, AND PLATO

1. Fideler, *Jesus Christ, Sun of God,* Quest 1993 for examples with geometry.

2. Richard Heath, *Sacred Geometry: Language of the Angels,* 115–20.

3. McGilchrist, *The Master and His Emissary.*

4. See my *Harmonic Origins of the World.*

5. "Metis," Wikipedia.

6. See my *Harmonic Origins of the World* or chapter 5 of *Sacred Geometry: Language of the Angels* for details of the planetary model.

7. See Reiche, "The Language of Archaic Astronomy," fig. 7, for an interpretation of Poseidon's Island.

8. McClain, *The Myth of Invariance.*

7. CAPPADOCIAN CROSSROADS IN ANATOLIA

1. Robin Heath, *Sun, Moon and Stonehenge,* 74.

2. Ouspensky, *In Search of the Miraculous,* 311.

3. Bennett, *The Masters of Wisdom.*

4. Bennett, *Creation,* 104.

5. Kostof, *Caves of God.*
6. Richard Heath, *Sacred Geometry: Language of the Angels,* 68–72.

8. THE ARCTIC ORIGINS OF ASTRONOMICAL SYMBOLISM

1. Cunliffe, *Europe between the Oceans.*
2. "Polar night," Wikipedia.
3. Rg Veda 1.164.

9. THE VEDAS IN SOUTHEAST ASIA

1. MacKie, *Science and Society in Prehistoric Britain.*
2. "Angkor: Asia's ancient 'Hydraulic City'" by Marissa Carruthers, December 2, 2021 on BBC, Travel blog.
3. See my *Harmonic Origins of the World* for a fuller picture.
4. See Bremer, *Plato's Ion.*
5. See my *Harmonic Origins of the World* for a full treatment.

10. ROME: THE GODDESS WITHIN THE VATICAN

1. See Wikipedia, "St. Peter's Basilica," section "Plan to rebuild."
2. I gave a fuller interpretation of the Cosmati pavement in my book *Sacred Geometry: Language of the Angels.*
3. "Vitruvian Man," Wikipedia.

BIBLIOGRAPHY

Allen, J. Romilly. *Early Christian Symbolism in Great Britain and Ireland: The Romano-British Period and Celtic Monuments with an Account of Early Christian Symbolism in Foreign Countries.* 1885. Reprint, Literary Licensing, LLC. 1992.

Bachenheimer, Avi. *Gobekli Tepe: An Introduction to the World's Oldest Temple.* Victoria, Australia: Birdwood, 2018.

Banou, Emilia. "Minoan 'Horns of Consecration' Revisited: A Symbol of Sun Worship in Palatial and Post-Palatial Crete?" *Mediterranean Archaeology and Archaeometry* 8, no. 1 (2008): 27–47.

Beck, Roger. *Planetary Gods and Planetary Orders in the Mysteries of Mithras.* New York: E. J. Brill, 1988.

Bennett, John G. *Dramatic Universe,* vol. 4, *History.* London: Hodder & Stoughton, 1965.

———. *Gurdjieff, Making a New World.* London: Touchstone, 1973.

———. *The Masters of Wisdom.* London: Touchstone, 1977.

———. *Creation: Studies from the Dramatic Universe Number Three.* London: Coombe Springs Press, 1978.

Berriman, A. E. *Historical Metrology.* London: J. M. Dent and Sons, 1953.

Bremer, John. *Plato's Ion: Philosophy as Performance: Philosophy as Peformance,* D & F Scott Publishing, Inc., 2006.

Brown, William Norman. *The Creation Myth of the Rig Veda.* 1942. *Journal of the American Oriental Society,* 62/2, 85–98.

———. *Agni, Sun, Sacrifice, and Vac.* 1965. *JAOS.* 88/2, 199–218.

———. *Theories of Creation in the Rig Veda.* 1965. *JAOS.* 85/1, 23–34.

Calasso, Roberto. *Literature and the Gods.* New York: Knopf, 2001.

Carrington, Philip. *The Primitive Christian Calendar: A Study in the Making of the Marcan Gospel.* London: Cambridge U.P., 1952.

Carruthers, Marissa. "Angkor: Asia's ancient 'Hydraulic City'" BBC Travel blog, 2021.
Colson, F. H. *The Week: An Essay on the Origin and Development of the Seven-Day Cycle.* Cambridge: Cambridge University Press, 1926.

Cunliffe, Barry. *Europe between the Oceans, 9000 BC–AD 1000.* London: Yale University Press, 2011.

Eratosthenes and Hyginus. *Constellation Myths, with Aratus's Phaenomena.* Translated by Robin Hard. Oxford: Oxford University Press, 2015.

Fideler, David. *Jesus Christ, Son of Gods.* Quest, 1993.

Frye, Northrop. "New Directions from Old." In *Myth and Mythmaking,* edited by Henry A. Murray. New York: George Braziller, 1960.

Giovannini, Luciano, ed. *Arts of Cappadocia.* Geneva: Nagel, 1971.

Graham, J. Walter. "The Minoan Unit of Length and Minoan Palace Planning." *American Journal of Archaeology* 64, no. 4 (October 1960): 335–41. Online: Archaeological Institute of America, stable URL: 618.

Graves, Robert. *The Greek Myths.* London: Penguin, 1955.

Gurdjieff, George I. *Beelzebub's Tales to His Grandson.* London: Routledge & Kegan Paul, 1950.

Haklay, Gil., and Avi Gopher. "Geometry and Architectural Planning at Göbekli Tepe, Turkey." *Cambridge Archaeological Journal* 30, no. 2 (2019): 343–57.

Heath, Richard. *Matrix of Creation.* Rochester, Vt.: Inner Traditions, 2002.

———. *Sacred Number and the Origins of Civilization.* Rochester, Vt.: Inner Traditions, 2007.

———. *Precessional Time and the Evolution of Consciousness.* Rochester, Vt.: Inner Traditions, 2011.

———. *Sacred Number and the Lords of Time: The Stone Age Invention of Science and Religion.* Rochester, Vt.: Inner Traditions, 2014.

———. *Harmonic Origins of the World.* Rochester, Vt.: Inner Traditions, 2018.

———. *Sacred Geometry: Language of the Angels.* Rochester, Vt.: Inner Traditions, 2021.

Heath, Robin. *Sun, Moon and Stonehenge.* Cardigan, UK: Bluestone Press, 1998.

———. *Bluestone Magic: A Guide to the Prehistoric Monuments of West Wales.* Cardigan, UK: Bluestone Press, 2012.

———. *Proto Stonehenge in Wales.* Cardigan, UK: Bluestone Press, 2014.

———. *Temple in the Hills.* Cardigan, UK: Bluestone Press, 2016.

Heath, Robin, and John Michell. *The Measure of Albion: The Lost Science of Prehistoric Britain.* Cardigan, UK: Bluestone Press, 2010.

Heath, Sir Thomas. *Aristarchus of Samos: The Ancient Copernicus.* Oxford: Clarendon, 1913. Reprint, New York: Dover, 1981.

Hesiod. *Theogony.* London: Penguin Group, 1976.

Kostof, Spiro. *Caves of God: Cappadocia and Its Churches.* New York: Oxford University Press, 1989.

Lewis, C. S. *The Discarded Image.* Cambridge: Cambridge University Press, 2013.

Lomsadalen, Tore. *Sky and Purpose in Prehistoric Malta: Sun, Moon, and Stars at the Temples of Mnajdra.* St. David, Wales: Sophia Centre Press, 2014.

Loveday, Roy. *Inscribed across the Landscape: The Cursus Enigma.* Gloucestershire, UK: Tempus, 2006.

MacKie, Euan. *Science and Society in Prehistoric Britain.* London: Paul Elek 1977.

Mannikka, Eleanor. *Angkor Wat: Time, Space and Kingship.* O'ahu: University of Hawai'i Press 1998.

McClain, Ernest. *The Myth of Invariance: The Origin of the Gods, Mathematics and Music from the Rg Veda to Plato.* York Beach, Me.: Nicolas-Hays, 1976.

———. *The Pythagorean Plato: Prelude to the Song Itself.* York Beach, Me.: Nicolas-Hays, 1978.

McGilchrist, David. The Master and His Emissary: Yale, U.P. 2009.

Michell, John. *Ancient Metrology.* Bristol, UK: Pentacle Press, 1981.

———. *At the Centre of the World.* London: Thames & Hudson. 1994.

———. *Dimensions of Paradise.* Rochester, Vt.: Inner Traditions, 2008.

———. *Twelve-Tribe Nations.* Rochester, Vt.: Inner Traditions, 2008.

Michell, George. *The Hindu Temple: An Introduction to Its Meaning and Forms.* Chicago: University of Chicago Press, 1988.

Nafilyan, Guy. *Angkor Wat: description graphique du temple. Mémoire archéologique IV.* Paris: École française d'Extreme-Orient, 1969.

Neal, John. *All Done with Mirrors: An Exploration of Measure, Proportion, Ratio and Number.* London: Secret Academy, 2000.

———. *Ancient Metrology,* vol. 1, *A Numerical Code: Metrological Continuity in Neolithic, Bronze, and Iron Age Europe.* Glastonbury, UK: Squeeze, 2016.

———. *Ancient Metrology,* vol. 2, *The Geographic Correlation: Arabian, Egyptian, and Chinese Metrology.* Glastonbury, UK: Squeeze, 2017.

Nicolas, Antonio T. de. *Meditations Through the RgVeda: Four-Dimensional Man.* Stoney Brook, N.Y. State: Nicolas-Hays, 1976

Ouspensky, P. D. *In Search of the Miraculous.* London: Routledge & Kegan Paul, 1950.

Paulsson, Bettina Schultz. *Time and Stone: The Emergence and Development of Megaliths and Megalithic Societies in Europe.* Oxford: Archaeopress, 2017.

Pendlebury, J. D. S., *The Archaeology of Crete.* New York: Norton 1965.

Petrie, W. M. Flinders. *Inductive Metrology; or, The Recovery of Ancient Measures from the Monuments.* 1877. Reprint, Cambridge: Cambridge University Press, 2013.

Reiche, Harald A. T. "The Language of Archaic Astronomy: A Clue to the Atlantis Myth." In *Astronomy of the Ancients,* edited by Kenneth Brecher and Michael D. Feirtag. Cambridge, Mass.: MIT Press, 1980.

Richer, Jean. *Sacred Geography of the Ancient Greeks: Astrological Symbolism in Art,*

Architecture, and Landscape. Translated by Christine Rhone. Albany: State University of New York Press, 1994.

Rodley, Lyn. *Cave Monasteries of Byzantine Cappadocia.* London: Cambridge University Press, 1985.

Rogers, John H. "Origins of the Ancient Constellations: I. The Mesopotamian Traditions, and II. The Mediterranean Traditions." *Journal of the British Astronomical Association* 108, no. 1 (1998): 9–28, and no. 2 (1998): 79–89.

Rountree, Kathryn. "The Case of the Missing Goddess: Plurality, Power, and Prejudice in Reconstructions of Malta's Neolithic Past." *Journal of Feminist Studies in Religion* 19, no. 2 (Fall 2003): 25–43

Santillana, Georgio, and Hertha von Dechend. *Hamlet's Mill: An Essay on Myth and the Frame of Time.* Boston: Godine, 1977.

Schultz, Joachim. *Movement and Rhythms of the Stars.* Edinburgh: Floris, 1986.

Shah, Idries. *Seeker after Truth.* London: Octagon Press, 1982.

Shaw, Joseph W. "Evidence for the Minoan Tripartite Shrine." *American Journal of Archaeology* 82, no. 4 (1978): 429–48.

Spanuth, Jürgen. *Atlantis of the North.* London: Sidgwick and Jackson, 1979.

Sugiyama, Saboro. *The Archaeology of Measurement: Teotihuacan City Layout as a Cosmogram.* Cambridge: Cambridge University Press, 2010.

Tilak, B. G. *The Arctic Home of the Vedas.* Puna, India: Tilak Bros., 1903.

Teteriatnikov, Natalia B. *The Liturgical Planning of Byzantine Churches in Cappadocia.* Orientalia Christiana Analecta 252. Rome: Pontificio Istituto Orientale, 1996.

Thom, Alexander, and Archibald Stevenson Thom. *Megalithic Remains in Britain and Brittany.* Oxford: Clarendon Press, 1978.

Thomson, George. "The Greek Calendar." *Journal of Hellenic Studies* 63 (1943): 52–65.

———. *The Prehistoric Aegean: Studies in Ancient Greek Society.* 1949. Reprint, London: Lawrence and Wishart, 1978.

Tompkins, Peter. *Secrets of the Great Pyramid.* New York: Harper & Row, 1971.

Trump, D. H. *Malta: An Archaeological Guide.* 1979. 2nd ed., Valletta: Progress, 1990.

———. *Malta: Prehistory and Temples.* Malta: Midsea Books, 2002.

———. *The Prehistory of the Mediterranean.* London: Allen Lane (Penguin), 1980.

———. *Skorba: Excavations Carried Out on Behalf of the National Museum of Malta 1961–1963.* London: Oxford University Press, 1966.

Ulansey, David. *The Origins of the Mithraic Mysteries: Cosmology and Salvation in the Ancient World.* New York: Oxford University Press, 1989.

Warren, William F. *Paradise Found: The Cradle of the Human Race at the North Pole.* Boston: Houghton Mifflin, 1885. Reprint, London: Forgotten Books, 2008.

Whitfield, Peter. *The Mapping of the Heavens.* London: British Library, 1995.

Yukteswar, Swarmi Sri. *The Holy Science.* Los Angeles: Self-Realization Fellowship, 1949.

INDEX